PENGUIN BOOKS

THE CELEBRANT

Eric Rolfe Greenberg is a native New Yorker who attended the University of Wisconsin and New York University's School of the Arts. He has worked as a publicist for Columbia Pictures, and as public relations director of *National Lampoon*. Mr. Greenberg lives in New York City with his daughter, Elinor.

THE CELEBRANT

A NOVEL BY

Eric Rolfe Greenberg

PENGUIN BOOKS

PENGUIN BOOKS
Viking Penguin Inc., 40 West 23rd Street,
New York, New York 10010, U.S.A.
Penguin Books Ltd, Harmondsworth, Middlesex, England
Penguin Books Australia Ltd, Ringwood, Victoria, Australia
Penguin Books Canada Limited, 2801 John Street,
Markham, Ontario, Canada L3R 1B4
Penguin Books (N.Z.) Ltd, 182–190 Wairau Road,
Auckland 10, New Zealand

First published in the United States of America by Everest House 1983
Published in Penguin Books 1986

LIBRARY OF CONGRESS CATALOGING IN PUBLICATION DATA
Greenberg, Eric Rolfe.
The celebrant.
1. Mathewson, Christy, 1880–1925—Fiction.
I. Title.
PS3557.R3783C4 1986 813'.54 85-21826
ISBN 0 14 00.8746 1

Printed in the United States of America by
R. R. Donnelley & Sons Company, Harrisonburg, Virginia
Set in Granjon

for
Jacob Shapiro

HISTORICAL NOTE

THESE ARE the facts of Christy Mathewson's life and career: born, Factoryville, Pennsylvania, 1880; attended Bucknell College, 1898–1901; joined the New York Giants, 1900; pitched a no-hit game in St. Louis on July 15, 1901, and another in Chicago on June 13, 1905; threw three shut-outs in the 1905 World Series, a feat never since equaled; was the pitcher of no decision in the tie game of September 23, 1908 and the loser in that year's playoff, and in the finale of the 1912 World Series; retired from play in 1916 with a career record of 367 victories, 186 defeats, 77 shut-outs, 2502 strikeouts; managed the Cincinnati club, 1916–18, resigning in the wake of the Hal Chase scandal; enlisted in the U.S. Army, 1918, served in post-Armistice Europe and was exposed to poison gas, never recovering his health; coached briefly for the Giants, 1919, and watched the 1919 World Series at the side of reporter Hugh Fullerton, noting doubtful plays on Fullerton's scorecard; retired to Saranac Lake, New York, and died there, October 7, 1925. He was one of five original members elected to baseball's Hall of Fame at its founding in 1936.

The Celebrant

Be sad, as we would make ye. Think ye see
The very persons of our noble story
As they were living. Think you see them great,
And follow'd with the general throng, and sweat
Of thousand friends. Then, in a moment, see
How soon this mightiness meets misery.

Prologue,
HENRY VIII

July 15, 1901—at St. Louis

										R	H	E
NEW YORK	2	2	0	0	0	0	0	0	1	5	10	1
ST. LOUIS	0	0	0	0	0	0	0	0	0	0	0	1

BATTERIES: New York, Mathewson & Warner;
St. Louis, Sudhoff & Ryan.
Winning pitcher: Mathewson. *Losing pitcher*: Sudhoff.

STANDINGS OF THE CLUBS

	W	L	PCT.	GB
Pittsburgh	43	26	.623	—
St. Louis	40	31	.563	4
Philadelphia	37	31	.544	5½
New York	33	29	.532	6½
Brooklyn	37	34	.521	7
Boston	30	34	.469	10½
Cincinnati	30	39	.435	13
Chicago	24	50	.324	21½

ONE

OUR FAMILY came to New York in the winter of '89, and in the spring I saw my first game of baseball. I was eight. My mother's brother, a jeweler, had preceded us across the Atlantic; Uncle Sid's family was small—only four children—but the crowding was awful, and in April my father thought to find a place of our own across the river in Brooklyn. He inquired of a German-speaking landlord near Prospect Park, who asked if there were children; my father answered "nine," the German nodded, and the two men shook hands. The following Sunday we stacked all we owned onto a rented dray and, daunted by the spectacular height of the new bridge, crossed the river by ferry. We returned before nightfall, much abashed. The German had understood my father to say he had no children, and when he arrived with nine the door was slammed in his face.

I missed this famous encounter. I'd run off to the park after my older brother Eli and discovered there clay diamonds cut into an immense field of grass, an expanse so generous that half a dozen ballgames were underway at once, some so distant that they appeared contests between toy miniatures. But toy players could hardly have amazed me more than these—adults! grown men playing games!—and many in uniforms complete with gaudy striped stockings. Far away a bat struck a ball; seconds later I heard the sound. This phenomenon was a great excitement, and I ran back and forth altering the time lapse until I was nearly trampled by an outfielder in furious pursuit of a fly ball. I dodged as he sprang, and as he stretched out in full flight the side of his shoe caught my cheek. I reeled and fell down. The ball bounded on, and the man who'd kicked me spat dirt from his mouth, looked up, and said, "Well, *shit!*"——the first words ever spoken to me on a ballfield.

The family stayed in Manhattan. We found the game everywhere, in every imaginable variation. There were large lots along both rivers which allowed the full exercise, and each street and alley had its own rules and exceptions. First by imitation, then by practice, we learned the game and the ways of the boys who played it, the angle of their caps, the intonations of their curses and encouragements. Our accents disappeared, our strides became quick and confident. My lefthandedness, regarded by my parents as a devil's curse, turned to my advantage in the pitcher's box. I threw a submarine ball, my knuckles grazing the dirt as I released it. "Get those knuckles dirty, Jackie!" my infielders would shout—Jackie, not Yakov.

The new fashion of overhand pitching soon threatened my eminence. When the National League legalized the pitch the neighborhood clubs followed suit; I tried the style but had trouble keeping the ball low, and it's the high pitch that's hit a distance. Our team enlisted a big fellow who could throw the overhand pitch with mustard, and to counter his advantage I attempted to learn the curve. It was accounted a disreputable pitch, the refuge of a trickster who had not the honest strength to power the ball over the plate. I found it difficult, for my fingers were too short to give the ball sufficient spin. Soon I was in the outfield, not liking the change; pitching is the core of the game. Vowing to master the curve, I threw to Eli for an hour every evening in the narrow alley behind our tenement. The youngsters on the block would come out and carry on as if I were pitching for the Giants against the Orioles for the Temple Cup: "Keeler's up, Jackie, watch it now! Ball one—come on, put it over! The curve, the curve! Oh, he hit it! Base hit! Now Jennings—better bear down, Jackie! McGraw's on deck!"

Finally, at fifteen, I made the curve ball work. I threw it as hard as the fast one, and it broke just as it reached the plate, a small break but a sharp one, straight down. It was a ground ball pitch, and it kept my infielders busy. In tandem with the big fellow I won a good many ballgames and achieved something of a neighborhood celebrity.

At the close of the '97 season the league sponsored an awards dinner at a restaurant on Grand Street, and there I shook hands with a major league player. His name was Jack Warner. I'd watched him play shortstop for an uptown semiprofessional club, marveling at his size and power, but he lacked speed, and the professional leagues had made him first an outfielder, then a catcher. His face showed the marks of that ignoble position. At the dinner, his eyes never leaving the page, he delivered a speech analogizing baseball and life. Practice, dedication, clean living, and fair play—these guaranteed success on and off the field. We froze in a handshake while a photographer immortalized the moment, and then I took a silver cup from him. He turned to the organizer of the event and asked for his fee.

In the spring I was asked to a professional try-out at Manhattan Field, far uptown. Men with leather faces and tobacco-stained teeth examined me microscopically and stood in at bat as I pitched. A week later a letter arrived, the first I'd ever received, offering a contract with the Altoona club of the New York–Pennsylvania League.

My parents wouldn't hear of it.

Prizefighters, jockeys, ballplayers: these were professional athletes. Most celebrated were the fighters at the championship level; heavyweight Corbett could get a room at most hotels, and it was said that the better people of San Francisco welcomed him to their salons. The exemplary Gentleman Jim thus crossed once impregnable barriers, but the gulf between his status and that of the journeyman was vast. Club fighters were neighborhood heroes, yet the unspoken assumption was that they fought because they were unfit to do anything else; a man did not opt for the ring, he was condemned to it. Most jockeys were black, apprenticed to trainers or breeders; one thought of them as one thought of the horses they rode. Some professional ballplayers were locally bred, but an increasing number were itinerants from distant farmlands who lived out of cardboard suitcases in back street boarding houses. Annually they jumped from league to league and team to team for the sake of a few dollars increase in salary. To bankers and landlords and shopkeepers they were suspect, disreputable; to a man with daughters they were dangerous. An underclass supported them. Ballpark crowds were mean and roistering.

I was seventeen, and done with school; I'd stayed at it longer than most. All the pressure of a family's traditions, hopes, and plans pressed down upon me. I was the fifth son; the first had driven teams to put the second through college, and the third, Eli, sold my Uncle Sid's jewelry to see the fourth through his studies. I was obliged to provide for young Sam's higher education. I argued that I could do this on a ballplayer's salary, twenty dollars a month, but though my parents came to believe that I'd actually be paid to play ball the issue went far deeper than money. We had not crossed the ocean to find disgraceful employment.

I had not the wherewithal to resist my parents. Rather than be dead to them—for my father threatened to turn his back to me and say a kaddish for my soul—I put away the contract and assumed Eli's position at Uncle Sid's jewelry store, while Eli packed a case of samples and sought markets in faraway cities. I was no sales clerk, nor as proficient as my cousins in working the precious metals and stones, but I did show a flair for design. Should a customer find nothing on display to his liking, I'd inquire what he had in mind and quickly sketch a model; often enough, a sale would result. Once I turned this trick to delight a trio of sisters who were shopping for an anniversary present for their parents. When the piece was done I delivered it to their home, far uptown in Turtle Bay—and what a home, a mansion with fantastic wrought-iron fencing at the doors and windows and an interior elegance which to my mind quite belied the small *mezuzah* posted at the entry. That Jews could achieve such grandeur was well nigh unbelievable.

The next week the youngest of the daughters returned to the shop to thank me and to order another piece. She was merry; she kept me so long at the sketch pad with her suggestions and alterations that Uncle Sid had to bustle us out of the store at closing time. I escorted her home by trolley, and learned her name —Edith—while she learned of me everything I could babble in an hour. When our club played in Central Park the following Sunday she was there with her sisters; they carried matching parasols that spun prettily in their gloved hands. I pitched well that day, breaking off the curve ball time after time. Afterward we shared lemonade in the Ramble.

The next day I could hardly lift my arm. The curve ball was

proving too great a strain. I needed a week between turns, and soon that wasn't enough. In the end I was back in the outfield, where the dream of Altoona and the big leagues faded. Playing less, I found time for afternoons with Edith; often I had to choose between her company at a concert or museum and Eli's for a ballgame. Once my brother had offered to escort all three Sonnheim girls to the Polo Grounds; the suggestion scandalized Edith's father, who coldly opined that baseball had ceased to be a gentleman's game after the Civil War, when it had been taken over by professional athletes. Might as well recommend a tour of the Tenderloin!

In deep summer, with Edith at her family's lodge on Lake George, I was ever at the ballpark—the new Polo Grounds, where the Giants were declining from their glory years of the early 'Nineties, or more frequently the Atlantic Avenue park in Brooklyn to watch the Superbas, stocked with Baltimore veterans, driving to the championships of '99 and 1900. Eli would introduce me to his sporting friends as his "expert" and make a great show of consulting me in whispers before placing a bet. It seemed he required a wager to excite his interest, whereas for me the game was all.

As the spring of 1901 turned toward summer, Eli urged me to join him on his annual "big swing" west to the Mississippi, north to Chicago, and east along the Lakes. "Sport, we'll eat a steak and see a ballgame in every city!" he swore. Uncle Sid accepted Eli's contention that my designs would be improved with a better knowledge of our markets, and Edith was away. I bought a suitcase—leather, not cardboard—and packed for my first excursion beyond the Hudson. Our first calls were in Philadelphia, where the steaks weren't much to speak of and the ballgame worse. The National League club was on the road, and we decided to test the infant American League, which had no New York franchise. The great attraction was Napoleon Lajoie, whose batting average for the Americans was a hundred points above what he'd hit for Philadelphia's Nationals the year before. His outrageous success seemed proof of the new league's inferiority. Nap was one of several former collegians courted for the club by Philadelphia's gaunt manager, Connie Mack; another was Plank, who pitched that day and easily beat the Milwaukee entry. Lajoie had three hits and scored twice.

We caught up with the Philadelphia Nationals in Pittsburgh. They missed Lajoie; the home club beat them authoritatively to move further ahead in the race. In Cincinnati dwelt a cadet branch of the family, and we stayed nearly a week. A second cousin whose face was a startling female approximation of my brother Sam's took me on an outing along the river and somehow contrived a rainstorm, a sheltering toolshed, and a blanket, but she could not contrive a different face, and the matter came to nothing. To my joy, the Cincinnati Red Stockings arrived for the latter part of our visit, and with them our own wonderful Brooklyn club, Ned Hanlon and his boys, the defending world champions. But the club was fading, the old Baltimore heroes further past their prime, and we sensed there would be no pennant flying over Atlantic Park at year's end.

On a Saturday morning in mid-July we stepped off the west-bound train into the heat of St. Louis and proceeded to the Chase Hotel. My brother always put up at the best hotels, signing as "E. Kapp." The name Kapinski would not be welcome on those registers. As I unpacked he searched out sheaths of the hotel's stationery and addressed invitations to the buyers of the town, requesting the honor of their presence at our "suite" on the Monday following. That done, he summoned an assistant manager to lay plans for breakfast, mid-morning, and luncheon service: oysters and champagne, deviled eggs, a variety of sausage to suit the Swiss and Germans among our clientele. (Later Eli would cadge some kosher delicacies for our correligionists, some of whom would first set foot in the Chase at our invitation.) He would welcome them all; he would tell the latest tales and gags; he would profess that, had he a choice, he would visit them at their offices, but to carry his samples on the dangerous streets was to court dire attack; he would show his wares and write his orders. In such ways did Eli advance the family business as it increasingly turned from retailing to manufacture. A middle-range market was emerging between the jeweler's private customer and the mail-order public, and the avenue to it was the department store. Our product line was priced below the artisan's custom work and above the C.O.D. merchandise that Eli accurately described as cheap junk. His job was to secure outlets against the efforts of dozens of direct competitors, and he shared the rails with a gaggle of salesmen hawking hardware or hose or outrageous feathered wraps; together they were inventing the business of the new century. Not all stayed at the grand hotels, but my brother would have put up at the Chase (or Boston's St. James or Baltimore's Belvedere) had his suitcase carried sandstone or cement.

Yet for all this show, his delight was to invite guests to the local ballpark. He'd staked this out as his personal form of business entertainment, leaving the burlesques and bawdy houses to his fellows of the selling trade. Here professional calculation matched his personal inclination. For himself, Eli Kapinski of New York, no fancy out-of-town attraction could rival those of home, but a major league ballgame in any city carried a sort of guarantee. For his clients—what a disarming suggestion! An afternoon at the ballpark, so refreshing, so American!

Shortly before three o'clock on a cauterizing Monday we rented a hack, collected three buyers who'd breakfasted with us that morning, and set out for League Park. St. Louis in July was the hottest place on the circuit, the hottest place God ever made a city. Our guests blamed the heat for the team's sorry record in the National League; they'd never finished above fifth place. I pointed out that the city's entry had won four championships in the old American Association in the 'Eighties. "Mildest summers on record," I was assured. This season their club had known early success but now, beset by slump and injury, they were losing ground to Pittsburgh. Still the fans prayed for a cool summer and came to League Park in record number; more than twenty thousand, an unprecedented gathering, had encircled the field for the previous game of the current series with the New York Giants.

In the matter of rooting, a boy's first team is his team forever. I'd seen my first big-league game below Coogan's Bluff ten years before, and whatever Brooklyn's current success the Giants were the club of my heart, their championships my own. But seven years' famine had followed the feast, and a dozen managers had come and gone under the club's owner, Tammany politico Andrew Freedman. A great war was waged between Freedman and my own idol, Amos Rusie, the huge righthander who'd won thirty-six games in glorious '94. Rusie refused to sign a contract in '96, came back to win forty-eight games for a lesser club in '97 and '98, and then held out for two full seasons rather than throw another pitch for Freedman. He called the owner a liar, a chiseler, a welcher, and a cheat; Freedman called Rusie a Republican. They were both right. Without Rusie or any other player of great regard the Giants finished last in 1900. They'd replaced most of their infield and half their mound staff, but their current fifth place standing owed largely to the advent of a talented collegiate pitcher of their own, a righthander who was scheduled to throw for New York.

Play had just begun when we arrived at League Park. We watched from the outfield as the Giants mounted an early attack. Sudhoff, the elfin St. Louis pitcher, began badly: a walk, a base hit, and then a drive that skipped under the right fielder's glove and rolled to our feet. The tallest of our clients, a thin man with merry eyes, kicked the ball back toward the fielder. Far away, in the middle of the diamond, the umpire threw up his hands and shouted "Hold! Three bases!" Two Giants scored, and we moved with haste to the grandstand, a single-tiered wooden structure that rose along both foul lines. Having paid a quarter a head to enter the Park grounds, we were now charged as much again to gain the grandstand and needed fifty cents more for box seats beneath the low, shading roof. But an attendant at the box-seat turnstile swore there was no room for us in the shade, and when Eli protested he scolded us with Irish vigor: "You want to come out early these days, we're winning now, don't you know!" Finally we found room behind third base for three on one grandstand bench and for two more directly behind. The thin client and a bearded one flanked Eli while the stoutest squeezed in next to me. We doffed our jackets and loosened our collars. The Giants had scored no more, and the teams had changed sides. Little Jess Burkett stood in for St. Louis, hands high on the bat, feet spread wide. Behind him Patsy Donovan, who doubled as manager, picked at the tape around the handle of his bat. The sun burned down, the cries of the crowd floated in the humid air, and the umpire pointed a finger at the New York pitcher and bid him throw.

This pitcher was big—gigantic, compared to Wee Willie Sudhoff—yet his motion matched Sudhoff's for balance and ease until he pushed off the pitcher's slab with his right foot and drove at the plate with startling power. His follow-through ended in a light skip, and he finished on his toes, his feet well apart, his hands at the ready for a fielding play. I heard the umpire's call but didn't know if Burkett had swung and missed or taken the strike, for my eyes hadn't left the pitcher. He took the catcher's return throw and regained the hill in three strides. His broad shoulders and back tapered to a narrow waist; he wore his belt low on his hips, and his legs appeared taut and powerful beneath the billowing knickers of his uniform. A strong, muscular neck provided a solid trunk for a large head, and his cap was tipped rather far back on his forehead, revealing a handsome face and an edge of thick, light brown hair. He bent for his sign, rolled into his motion, and threw; this time I followed the ball and saw Burkett top it foul.

The fat man moved against me, reaching for his kerchief. "Who's your pitcher?" he asked.

"It's Mathewson," I said. "Christy Mathewson."

"That's Mathewson? Big kid!"

"He is that. He's bigger than Rusie, that's for sure."

"Throws hard."

"Yes, he does."

"He's winning for you, isn't he?"

"Eleven games, best on the club."

"Not bad for new corn."

"Actually, he pitched a few last year," I said.

"Win any?"

"No. As a matter of fact he lost two in relief. Then he went back to school."

Another fastball: Burkett's swing was late, and the ball bounced to first base. Ganzel gloved it and tossed underhand to Mathewson, who caught it in full stride and kicked the base for an out.

"He got over there in a hurry, too," the fat man observed.

"Hey, sport, take this and bring back some wieners and beer for us all, won't you?" said Eli, pushing a silver dollar into my hand. He winked and clapped my arm. "Hurry back now, I'm going to need your advice."

I struggled to the aisle and headed along the walkway that divided the grandstand benches from the box seats. As I reached the ramp I paused to watch Donovan at bat. His red face was lined with a manager's web of worry. Patsy fought off the pitches with short, choppy swings, hitting several foul before earning a base on balls. Ganzel met him at first, and they exchanged a greeting. Ganzel was ancient, nearly forty, and Donovan beside him looked as old. I scanned the Giants in the field: Strang and Hickman, on either side of second base, were no striplings, and Davis, New York's playing manager at third, was older yet. They were all past thirty in the outfield. The catcher was Jack Warner, he of the awards dinner years before; already he was a veteran of years. Schriver, ready at the bat for St. Louis—enough to say that they called Schriver "Old Pop." At the center stood Mathewson, young as an April morning in that sweltering July, and I, small in the crowd at the top of the ramp, turned and walked down into the shadow beneath the grandstand.

The vendors at their sizzling grills cursed loudly as fat spattered on their aprons. All about, sports in checkered vests argued, passed money, and wrote betting slips. Every inning, sometimes every pitch, was worth a wager. With my dollar I bought five pigs-in-a-blanket and as many bottles of beer, and pocketed two bits in change. I worked my way back up the ramp with some difficulty. Near our seats the fans came to my rescue, passing the food hand over hand to Eli with the efficiency of a fire brigade. I couldn't resist an urge to toss the quarter to my brother, who snared it backhanded. The section resounded with cheers, and the ballplayers on the field turned at the commotion.

"Still two-nothing, sport. Donovan was caught trying to steal. Hey, here we go again!" Eli cried as a base hit began the Giant second. But their game was all sock-and-run; after Strang's single, Warner flied out.

"They don't believe in the bunt, do they?" said the fat man.

"They won't win 'til they learn how," said Eli. "You've got to be able to lay it down, right, sport?"

"Here's Mathewson," I said. "He'll be bunting."

Eli looked at me. "A dollar says he brings off the sacrifice. All around?"

The clients accepted the wager. Sudhoff pitched, and Mathewson pushed the ball onto the grass and ran to first with the sprightly grace of a smaller man. He was narrowly out.

"He can bunt," the thin man conceded.

"Double or nothing that they score," said Eli.

Sudhoff walked a man, and the next nailed the first pitch on a low line over second base. Two Giants crossed home.

"Double or nothing they score again!" said Eli. I pushed my knee into his back, but he looked over his shoulder and winked. The three buyers took the bet and cheered when a fly ball ended the inning and wiped out their debts. Mathewson walked slowly to the pitcher's mound, dug at the slab with his toe, smoothed the dust, and worked into his warm-ups. Again his size and youth impressed. The bearded man beside Eli studied him.

"Mathewson, his name is?"

"College kid," said Eli. "Connie Mack signed him for Philly, but he jumped to New York."

"Where's he from?"

"Pennsylvania," said Eli, at the same time that I said, "Bucknell College."

"Imagine a college man playing ball for a living!"

I mentioned Lajoie and Plank, whom we'd seen in Philadelphia, but the bearded man snorted that the new league would sign anyone, and while he knew Lajoie had played in a college uniform he doubted the boy had ever seen the inside of a classroom. Nor had Mathewson, he'd wager.

The fat man turned to me. "But you said he went back to school after last season!"

"Quit the team in September to do it," I said.

"You've seen him before?"

No, I explained; in recent seasons the more talented Brooklyn club had caught my fancy, and I hadn't been among the few at the Polo Grounds when the rookie threw his first big-league pitches that summer or won his first victories in the spring. What I knew of Mathewson came from notes in the newspapers: his age, a year greater than my own, and his home, the farming country of the Susquehanna Valley; his fame as a Bucknell footballer. I knew he'd pitched in professional leagues in New England and Virginia, and that while he'd put his name to a major league contract with Connie Mack he'd never worn a Philadelphia uniform. Instead he'd come to New York, and now, in his first full year, he had a third of the club's victories. If his skills were the test he belonged in the National League, but like the bearded client I wondered why a true collegian would choose the life of a professional ballplayer.

The second inning ended quickly, and while Sudhoff hit his stride in the third Eli began to orchestrate wagers with every batter. If I thought the proposition doubtful I'd signal by pressing my knee into his back, but he ignored my advice as often as he accepted it, and I came to understand that a deeper game was in progress. Eli was selling jewelry, and it wouldn't do to take too much of his clients' money. Each time he won he offered double or nothing on the next bet; at best he would finish little better than even, and at worst far worse. I engaged in chatter with the partisans around us when the Giants batted, or gazed over the midwestern crowd dotted with wide-brimmed western hats among the standard derbies and occasional boaters. Far down the right field line was the only uncrowded grandstand section; there the coloreds sat in overalls and yellow straw hats. When St. Louis batted I studied Mathewson. He could throw hard, and with excellent control, shading the edges of the strike zone, mixing his fastball with a curve that seemed somehow erratic. Sometimes it fell an astonishing measure, while other times—what did it do? Certainly it behaved differently from the drop, but from our location I couldn't track the pitch. When St. Louis came to bat in the fifth inning I excused myself and threaded my way to a spot close behind home plate. Now I could see the inner game, the fierce battle between pitcher and batter where power and control sought mastery over instinct and guess.

Mathewson began against Padden with a pitch that came in

hard at belt level and dropped abruptly and dramatically, a superb overhand curve, and one which had wrenched my arm when I'd tried it. The second pitch was a fastball on the outside part of the plate, a second strike. Now another breaking ball, but so unlike the first, slower, and breaking in reverse, in the nature of a left-hander's curve; I'd never seen a righthander throw one. Nor had Padden, who swung late and was lucky to tip it foul. Padden stepped out of the box and shook his head like a man who'd just seen a rabbit leap out of his own hat. He took his stance a bit closer to the plate, leaning over to guard the outside corner. Then in an instant he was on his back in the dirt; comically, the bat landed on his head. Mathewson's fastball had reclaimed that disputed inside territory. Padden dusted off his knickers and took his stance farther off the plate. Mathewson stretched and threw another reverse curve, and Padden missed it badly. Strike three.

Mathewson worked through the St. Louis lineup just so. Always the first pitch was a strike, and usually the second; then a teasing pitch down low, or that strange fading curve thrown where no batter could hit it squarely, if at all. His rhythm and motion were balletic: a high kick, a swing of the hips, a stride forward, and finally the explosive release of the ball. There was intelligence as well as power behind the pitches: he had a four-run lead and found it to his purpose to walk Kruger with two out in the fifth and then retire the light-hitting Ryan, and to pass the dangerous Burkett in the sixth in favor of facing Donovan, who grounded out. When I rejoined Eli in the seventh inning Mathewson had allowed three baserunners, but no hits.

"Think they'll get to him, sport?"

I'd never witnessed a no-hit game. I'd come close to pitching one four years before, as close as the eighth inning, but a swinging bunt that squiggled up the first base line had foiled me, and in my disappointment I'd been racked for three runs. "He's doing pretty much what he wants," I said.

"A dollar says he gets by Schriver," Eli offered; the odds were heavily against a no-hitter, and betting with Mathewson seemed the safest way to protect the clients. Old Pop took a fastball for a strike, and the crowd booed. Its cries had taken on the anger of the heat and the temper of frustration. The fat man ground his cigar beneath his patent leather boot and muttered something about Schriver; the club missed its injured sluggers. Mathewson pitched, and Pop grounded weakly to Strang at second base. One out in the seventh.

"Double or nothing, all around?"

Now Padden again, and the first pitch a high strike, the second higher yet—but Padden had no wish to wait for another curve and swung, lifting a fly to center field where Van Haltren had hardly to move to glove it. Two outs now.

"Double or nothing?"

"It's a bet."

"What happens on a walk?" I asked.

"Do you think he'll pass him, sport?"

"He'll be careful with Wallace. It's his pattern."

Eli nodded. "I say he'll get him out. Double or nothing on an out." Mathewson's fastball flashed, and Wallace took it and jawed loudly at the umpire's strike call. Now that strange breaking pitch, and Wallace bounced it to third; he was out by two steps at first. The Giants came off the field as the crowd booed all the more, at Wallace, at the umpire, at the summer's merciless heat, at the cast of impending defeat.

The fat man tapped Eli on the shoulder. "It's four dollars now, right?"

"Two," said Eli.

"Four."

"No, the bet on Schriver brought us even."

"We were even before then. It's four."

"Well, if you insist," said Eli, laughing.

The Giants went down swiftly in the eighth, but New York's efforts at bat hardly mattered now; we wanted Mathewson pitching. In the St. Louis half he began with a strikeout, and the buyer's debts to Eli doubled to eight dollars apiece. The total was nearly a month's salary to me. Next was Ryan, hardly a threat; after taking two strikes he chased a high fastball and lifted it to the center fielder's range. Van Haltren loped easily across the grass and reached out to cradle it, but the web of his glove seemed to fail him, and the ball dropped at his feet.

Twenty thousand roared, but the fat man shook his head. "An error," he said.

"It's all the same," said Eli, attempting a mournful expression. "We're even."

"No, as far as I'm concerned, it's an out."

"The man's on second base," said Eli.

"But the bet is on a no-hit game, isn't it? The pitcher against the hitter. The man should have been retired. Now, double or nothing?"

It seemed it was a matter of integrity, for which the fat man had a reputation; when Eli made to forgive the debts of the men flanking him they proved no less upright than their colleague. They could play the inner game as well as my brother, for what might be lost in money was more than made in future favor. Word spread from the press benches that "error" was indeed the official call; the no-hitter was intact. Eli shifted ground: why not call the play "no bet"? The compromise was accepted, and the balance among the men reverted to eight dollars apiece.

While the party debated I studied Mathewson. He showed no annoyance in the wake of the misplay. His team still led by four runs, the inning was late, and the weak-hitting Nichols was at bat with pitcher Sudhoff to follow. Mathewson set, glanced at the baserunner, and threw. Nichols' bunt was a surprise; the score, the inning, the pitcher on deck all argued against it. Catcher Warner was slow to move after the ball, but Mathewson was upon it instantly, the ball in his glove, then in his hand, then at first base. Two outs, and sixteen dollars due to Eli from each of his clients.

"Double or nothing?"

Sudhoff, the little pitcher, hit the ball sharply, but Mathewson snatched it out of the air and the inning was over. At the Giant bench Mathewson greeted center fielder Van Haltren with a forgiving slap on the rump. He was far more cheerful than Eli, a man due ninety-six dollars of easy money. When Mathewson batted in the Giant ninth he was applauded by the home fans, and I cheered. Mathewson tugged at the bill of his cap in acknowledgment. "Imagine, a college kid!" said the bearded client. Most of the crowd had conceded the game, and many would prefer a defeat of special regard to a spoiling single in the home ninth. For myself, I wanted three clean outs and a glorious end; our clients' purses were of no matter.

"Double or nothing?"

Burkett led off. Mathewson started with a fastball; for the hundredth time he took Warner's return throw and climbed the mound. On the right knee of his knickers was a round smudge of red clay, the emblem of a hundred strides and a hundred pitches launched. He bent, stretched, and pitched again, a fastball in on Burkett's hands; it ticked the bat and sailed past Warner, who walked slowly in the heat to retrieve it.

"Oh, college boy! Oh, you college boy!"

Mathewson stood on the hill waiting for the catcher to return to position, his left leg slightly bent, his weight on the right. He gloved Warner's toss and bent for his sign, the ball resting in the pocket of his small brown glove. He seemed as fresh as when he had begun, and quite still: no heaving breath, no sleeve drawn across his face to clear the summer sweat. All question left me. This was Mathewson's place and moment; my whole being was with him. Burkett would not deny him, nor Donovan next, nor old Pop Schriver, that dark moving figure on the St. Louis bench.

"So young!" muttered the bearded man.

And I was old, I thought. I was older than Mathewson, older than Schriver, older than any of them in uniform. My youth had ended on a ragged lot by the Hudson when the curve ball had beaten my arm and my spirit—no, when I'd folded the contract into a drawer and reported for work at Uncle Sid's shop. I was on the road, yes, but as an old man, hawking samples in old men's hotels, learning how I might bet to keep old men happy. I watched Mathewson, and he became my youth; it was my fastball burning by Burkett, it was my curve that little Jess lifted to the outfield, and after the ball came back and around the infield I felt it was my glove closing around it, my arm that launched the fastball at Donovan's knees and the next that cut the black of the plate on the outside. My youth made him chase a breaking ball in the dirt, and there were two outs; here was Old Pop, and I had the game and the no-hitter in my hand.

Curve ball: Schriver lets it pass for a strike.

The other, fading curve: Schriver, off balance, swings and ticks the ball foul. Ganzel picks it up barehanded and throws it to me. Schriver is nervous; I see his hands moving on the bat, his heel twisting in the dirt.

I waste a pitch high, ball one. I take Warner's toss, wrap the ball in my glove, jam the package under my arm, reach for the resin, dust my hand. There is a lone incomprehensible cry from the grandstand, then silence. I turn, I bend, I look for Warner's sign. I toe the slab. I stretch. I throw.

Ground ball.

I reach; it is past me—but Strang is there, he takes it on a high bounce, he waits for Ganzel, old Ganzel, to set himself at first, he snaps a sidearm throw, and the ball disappears into Ganzel's mitt.

The crowd's hoarse voice rises in the heat, the Giant bench empties, the fielders race to the mound, and the team leaps to touch and embrace——

Mathewson.

I left Eli stretched out upon the great bed in our room at the Chase, corpselike, his face covered with a white towel soaked in icewater. He was seven hundred and sixty-eight dollars the richer and had never felt worse. Before him lay a duty that ran counter to his taste and practice: his clients must greet the dawn with a sense of having won, not lost, two hundred and fifty-odd dollars apiece, and for this he must plow his winnings back into an evening of splendid debauch. There was about it, he admitted, an aspect of challenge: so vast a sum, so little time—and in St. Louis! (Though not St. Louis proper; he knew of a famous fancy house across the river, and with foursquare application he might lose much of the sum at poker, and with the rest purchase such pleasures as would please his partners.) Yet it was not E. Kapp's way of doing business, and, moreover, he must now entrust to me the preliminary work in Chicago, richest of all western markets. As I packed he repeated his instructions and made me recite them, twice: the number of our compartment on the overnight train, the name of the assistant manager at the

Palmer House, the whole list of buyers I must invite to our suite over the three days he'd allotted to Chicago, the exact wording of the invitations, the menu for each entertainment. He made me write his name dozens of times until I achieved an acceptable forgery. As I departed I offered this comfort: had the betting begun with the game's first batter, Mathewson's no-hitter would have exacted sixty-seven million dollars—and change—from each client. Eli shuddered.

I took a hack to the Union Station. The somewhat comic aspect of my brother's situation only briefly alleviated my own anxiety at going it alone to lay the groundwork in Chicago. Still, the prospect dismayed me less than the thought of accompanying Eli across the river that night. What a straitlaced pair we were, I thought; what little reputation would salesmen have for revelry if they were all cut with the die that had stamped out the Kapinski brothers! The Kapp brothers.

The train crossed the Mississippi at dusk and headed north along its eastern bank. An hour later there was nothing but dark fields under an indigo sky. The romance of travel had evaporated for me somewhere between Pittsburgh and Cincinnati, and the heat of this slow night did nothing to revive it. I sat alone in the compartment writing out drafts of invitations. It was too hot for appetite; I had the porter fetch some iced tea. He reported that there was a high time to be had in the parlor car, but I wanted no such distraction. I recited Eli's instructions for the fiftieth time and imagined all that might go wrong. At each stop I worried that some disaster had befallen us, that I was bewrecked and stranded and utterly helpless.

When the porter knocked to make up my bed he suggested that I have a bite to eat before the kitchen closed. He was so solicitous that I felt I must obey; I donned my jacket and left him to his task. As I entered the last coach before the parlor car I heard the sounds of raucous festivity. I'd witnessed some jolly affairs among the salesmen in our travels, but none so loud, so late at night. I stopped to watch through the door's glass window.

I recognized one of the celebrants immediately: Jack Warner, wrestling with a teammate over a pitcher of beer. I saw Van Haltren shelling peanuts and flipping them into his mouth, manager Davis pushing coins over a tabletop and explaining the patterns to a couple of younger players. I knew that the club's next games were in Chicago, but it had never occurred to me that the Giants would be aboard this train. There were two dozen in all, some on their feet sparring and roughhousing, others lounging about the car. Their sporting jackets were spotted with great stains of sweat and spilled beverage. Two stood in the center, arms about each other's shoulders, warbling with slight attention to harmony and much to expression. They were serenading Mathewson.

He stood with one foot on a chair, his white linen jacket carefully folded over his knee. Even in this posture he was yet taller than the songsters, by far the largest man in the car and, in that company of athletes, the most perfectly formed. His face was made of straight lines: a strong jaw, high cheekbones, a direct nose, a level brow. His eyes were intelligent and alert. His skin was deeply tanned. He was grinning, and when the song was done he threw his head back and laughed while the whole car rocked with applause and shouts.

I watched him through my own mirrored image in the glass and sensed an immense distance from him. He was everything I was not. I couldn't approach him; what might I say? I could gush in praise; it would be forgotten in an hour. I could talk of my own disappointed career; could he possibly sympathize, he who had mastered the craft I'd abandoned? There was a gulf between us that I felt I must not cross. I had nothing to offer him.

I turned and walked back to my compartment. I took out my drawing pad and my kit of styluses and colored inks and began a sketch. I had in mind a commemorative of the no-hit game, a ring. I drew a square center stone set diagonally in the geometry of the playing field and bordered it with four diamond baguettes. I set it on a shank of gold, and on the base I drew words for engraving: *St. Louis/ July 15, 1901.* I thought a faceted emerald would be fitting for the gemstone; the game is played on a field of gold and green. I colored the drawing roughly and examined it. It would do. I tried to design the etching for the band but the rolling train made the close work difficult, so I closed my kit and put the pad on the bed. I undressed, donned my robe, and went to the washroom. When I returned I sat on the bed and studied the sketch. The design was strong—I could think of nothing specific to improve it—yet the overall effect lacked vibrancy. I put the pad on the floor, doused the gas lamp, and thought on it. Soon I slept.

It was still dark when I awoke. The sheets of the bed were damp with sweat. I turned the pillow and lay wakefully, feeling the sway of the train and the dankness of the air. I wondered if I'd ever feel cool again, or if the whole journey would be framed in heat. Then I thought of the ring. I lit the lamp, reached for the pad and stylus kit, and began a new drawing. I retained the design in all its elements, but in place of the emerald I colored a polished ruby, red with fire.

Chicago's Palmer House was all burnished wood and potted palms, crystal chandeliers and fringed oriental rugs. The clerk held a cable for me from Eli: ARRIVE FIVE O'CLOCK DO EVERYTHING I SAID AND NOTHING ELSE. Ten words exactly. He'd survived the night. Through the morning I wrote and addressed invitations. The hotel saw to their delivery by hand. I lunched alone in my room; afterwards I made a finished drawing of the ruby ring, tore it from the pad, and put it in my pocket. At the desk I asked if there were a ballgame that afternoon; the desk clerk had no idea but a bellhop informed me that the National League game began at four at the West Side Grounds at Lincoln and Polk Streets. He added in a tone of authority that the home club was on the rocks and the visiting New Yorkers probably worse; they played a better brand of ball near the stockyards, where a pair of meat-packing plants were locked in bitter rivalry. I'd miss Eli's arrival if I attended the Giants' game, so I returned to my room.

Oh, my brother! He slogged in at five-thirty, taking slow, careful steps, and he winced at the noise when the bellhop dropped his suitcases in the wardrobe. "Pay him," Eli whispered, and I gave the boy a dime. Eli jumped as the door slammed shut.

"Is everything done?" he croaked.

"Yes."

"Good. I'm going to die." He fell onto the bed. I undressed him, covered him, shut off the light and left. I had dinner in the hotel restaurant, enjoying the steak, but the bill so shocked me that I forgot myself and signed my full name. I had to cross out the latter part and insert an extra P. Two-and-a-half dollars, indeed!

It was six-thirty, still quite light outside, and I thought to take some air. As I circled the block I saw the bellhop who'd been so informative about baseball in the town emerging from an alley. I nodded to him, and he asked if I'd caught the game that afternoon; I said I hadn't, but that I'd been in St. Louis to witness Mathewson's no-hitter.

"He's the real McCoy," the bellhop said. "The New Yorks win when he pitches and lose when he don't. Cubbies beat 'em today, with Waddell."

"Waddell! He was with Pittsburgh not three weeks ago. I didn't know he'd been traded."

"Hell, with the new league no one knows who's gonna be with who from one day to another. They're trading everyone who might jump, figure they'll get something for 'em while they can. Ole Rube won't be with us long."

"Have you seen the new league?"

"New league!" He snorted. "It's just the old Western League with a couple of big-city franchises. You know how he did it, this guy Ban Johnson who runs it? He blew into town one day last winter and looked up Clark Griffith, our best pitcher, touched 'im with a wand and says, presto! you're the manager of my new Chicago club. So, presto! the Cubs are on their ass and Griffith's in first place with the Americans. But it won't last. All the guys who jumped will just jump back next year. The new league is just too . . . too new, you know what I mean? You gotta grow up with a club to really care about it. I been a Cub fan since I was this high."

"I cut my teeth on the Giants."

"Yeah, they had some good clubs back then. I seen a lot of 'em. They usta stop at the Newton House, right near where I live. Near the ballpark."

"Do they still stay there?"

"Far as I know. Hey!" He called to a colleague in the alley. "You know if the New Yorks still put up at the Newton?"

"Yeah, they're there now."

"You see?" said the bellhop. "Tradition. Same clubs, same hotels, same as always. That's the ticket. The American League won't last another year. Too new."

I thanked him, and when he left with his friend I returned to the lobby to ask the desk clerk for directions to the Newton House. These he gave, but with a warning about that part of the city. It could be rough there at night, he said; was I sure it was the Newton House I wanted? I walked to Lincoln Street and boarded a trolley at a busy intersection. The route was a study in decline: fewer shops, fewer streetlights, fewer people. The black tar walls of the West Side Grounds' empty grandstand seemed forbidding. Two blocks beyond the ballpark I debarked, checked the street signs, and walked north.

Of hotels I knew only the grandest. Newton House was less than that. A tattered awning extended from a narrow doorway. Inside there was a small cubicle barred like a teller's cage, and in it a man stood on a desk, tinkering with a three-bladed fan above his head. "Two bits for a room, four for a room and a toilet," he said without looking at me. I replied that I was there to meet someone, and he shrugged. I stood in the threadbare hallway, feeling the occasional glances of the clerk, until I heard footsteps on the stairs above me. A derbied man carrying a *Police Gazette* appeared and stopped to watch the clerk in the cage. I recognized the Giants' catcher.

"You're Jack Warner," I said.

"Right. You looking for me?"

"Actually, I wanted to see Christy Mathewson."

"Matty ain't here. You a friend of his? College man?"

"No, I'm traveling on business. I'm a New Yorker."

"What part of town?"

"Allen Street."

"Jew?"

"Yes."

"I was born on Water Street. Live up on the Heights now."

"Near the ballpark."

"That's me. Walk to work with a lunchpail." He studied me, then stepped back and pointed. "You're a ballplayer, ain't you?"

I was astonished at the recognition and said as much.

"No, I remember. I gave you a trophy once at some dinner, got ten bucks for showing up. Jew food, I recollect. You ever sign with a club?"

"No, my arm went bad."

"Damn shame. What do you want with Matty?"

"I have something I wanted to show him."

"He might be at Riggs'."

"Riggs'?"

"It's down the block. Come on."

On the walk we traded what we knew of opponents in the city's leagues. Warner delighted to list those of his former teammates now in jail, and there were many; many more walked to work with a lunchpail. Only Warner played big-league ball. Riggs', a neighborhood tavern boasting a tile floor and a mirrored bar, was moderately crowded. Warner lined up for a cut of corned beef and cabbage while I bought two nickel beers. We looked for an empty table. "Hey, Dixie!" Warner shouted, and a young man in a straw boater waved from a corner spot. We made our way to him. It was Strang, the second baseman.

"Meet a dead-arm pitcher," said Warner, and I realized I hadn't given my name. "Jackie Kapp," I said as I shook Strang's hand.

"He's looking for Matty," said Warner.

"Matty's gone to the symphony," said Strang, pronouncing it with irony: sym-foe-nee. "Or maybe it was the mew-zee-um, or the oh-per-ah."

Warner laughed. "Dixie, the way he's going I'll buy him a ticket to every symphony on the circuit."

"Shit," said Strang.

"You're just sore 'cause you wanted to try to get even at checkers. He's been picking your pocket all year long."

"Well, I'm playing with a handicap, y'understand," said Strang. "He's smart, and I'm dumb."

"Are you ever! Why weren't you running in the eighth today? I hit the damn ball behind you and they still got you at second."

"And you at first. Shit, you're slower than a three-legged mule."

"I can move when I have to. Like out of Bertha's that time in Louisville—"

"When that fat-assed deputy came up those stairs—"

"Hollering he'd paid for her in advance—"

"And no numb-nuts ballplayer was gonna get in there 'fore he did!" They laughed together.

"What was her name?" said Strang.

"Louise."

"Louise. Shit, she was a ballplayer's ballplayer."

"Yes, she was. That was my first full year in the big leagues," Warner said to me, "and this horse's ass was a dumb rookie."

"All rookies is dumb," said Strang, " 'cept if they go to the sym-foe-nee."

"What did you want to show Matty?" Warner asked.

"A design of a ring. I'm a jeweler. That is, I design jewelry."

"You got watches?" said Strang. "A pocket watch? I mean, a really good pocket watch?"

"They're not in our line."

"Oh. I thought you could get me a price."

"I'll see what I can do."

"I'd 'preciate it."

"What kind of ring?" Warner asked.

I took the drawing from my pocket. "I saw the no-hit game in St. Louis," I said, "and I drew this. I wanted to show it to him."

"What about me?" said Strang as Warner studied the sketch. "I got him the last damn out. A tricky little grounder. That fadeaway comes off the bat with a crazy spin."

"I've never seen a pitch like it," I said.

"Neither has anyone else," said Warner, handing the drawing to Strang. "Jesus, he was throwing it right where he wanted to yesterday. He was never better with that pitch."

"And you're the guy who told him to forget about the fade-away!" said Strang. "I don't know what you're more, dumb or slow."

"What I told him was that with his fastball and his curve he didn't need a trick pitch, he could get by on the two. Hell, he could get by with one. With three, he's damn near illegal."

"How much?" said Strang, handing me the design.

"If it were actually fashioned? About two hundred dollars."

Strang whistled. "Pretty steep."

"Matty could afford it," said Warner. "He's got a topcoat worth that much."

"I wasn't going to ask him to buy it," I said.

"You're gonna give it to him?" said Strang. "Shit, the rich get richer and the poor get drunk."

I was embarrassed. "I hadn't thought about it," I said. "I did the drawing on the train last night. I simply wanted to show it to him."

"I don't understand," said Warner. "You just want to give him the drawing?"

"He wants to give Matty the drawing and have him order the ring," said Strang, grinning. "Pretty slick game. I guess you're a college man, too."

"He's a Jew," said Warner.

"Oh!" said Strang, putting both hands around his stein.

"I had no intention of asking Mathewson to buy the ring," I said. "My brother won seven hundred and fifty dollars on the game—"

"Oh! Well, then, Matty's damn well earned the ring, ain't he?" said Strang. "You've got some brass, trying to sell him—"

"Look, I'm sorry," I said, rising from the table. "It's a misunderstanding, really."

"I don't think so," said Strang.

"I'd better go." I took the sketch off the table.

"Listen, Jew," said Strang, "stay away from Matty, you hear me? Stay away from him!"

It was nearly dark, befitting my mood. I walked back to Lincoln Street and waited nervously for a trolley. The more I thought of the episode with Warner and Strang the more confused I became. I considered waiting for Mathewson at the Newton House, but the risk of encountering the two veterans again was more than I could face. I realized that I wanted Eli's advice. His counsel, however, would have to wait until morning, for he was snoring heavily when I returned to our room. I sat in a chair and looked out over the busy Chicago night, and I bit my lip and fought back tears.

Eli shook me awake. It was light; I was still in the chair, fully dressed. His concern for me was real, but interwoven with his anxiety about breakfast arrangements for the first contingent of Chicago buyers. Our conversation jumped jaggedly between the two problems and was exasperated by our sharing of the bath and wardrobe and the frequent interruptions by service staff as they arrived to prepare our room. I managed to tell him about the encounter with the Giant team on the train and the fiasco with Warner and Strang. Eli cursed the ballplayers and hugged me; the hotel detective entered to find us in embrace, and blustered. Suddenly all Eli's attention was on business. The assistant manager looked in, and he and Eli praised my success at this baptism; all was ready, nothing overlooked. When the buyers began to arrive I breathed a prayer of thanks and retired to a corner to admire Eli's recuperative powers and inimitable style. Two additional meetings followed the first, and Eli had appointments after luncheon; not until late afternoon, over coffee in the hotel restaurant, had we time to ourselves. It was then that I showed him the drawing of the ruby ring. He studied the design carefully.

"It's the best thing you've ever done," he said.

"I know. That's why I'm so upset."

"I wish we'd been together. We'd have bought a round for the whole club." His face darkened. "I wish I'd been anywhere but East St. Louis."

"What was it like?"

"Don't ask me about it, Jackie. I'd like to forget the whole thing." He examined the drawing. "What a talent you have, sport. But as a salesman . . ." He shook his head. "You don't have any idea what you did wrong with Warner and—who was the other one?"

"Dixie Strang."

"Strang. Good glove. You went wrong because you had no idea of what you were about, sport. You've got to know what you're after in a deal."

"I wasn't selling anything, Eli."

"I know that, but they didn't. How could they? Mean as they were, they were right to have their defenses up. It's the oldest game in the world to pretend you don't know what you're about and let the other fellow propose the deal you're trying to turn. I've used that trick a few times myself." He smiled. "You might be pulling it on me, for all I know."

"What do you mean?"

"You want us to make the ring and give it to Mathewson, isn't that so?"

"Could we?"

"Pretty expensive tribute. Can you give me a reason we should?"

"Because he's marvelous, Eli! Didn't you see him? Every first pitch a strike, every lead-off hitter set down! After the first inning every base on balls came with two out and brought up a lesser hitter. I've never seen anyone pitch a game like that!"

"How about Rusie?"

"Rusie was all smoke. Sure he could throw, but Mathewson pitches. He's marvelous."

"A bit too marvelous for my blood. Better for me he stayed at Bucknell."

"That's another thing, Eli. He doesn't even belong on a baseball field, a college man with money of his own. He could be a doctor or lawyer or stockbroker, but there he is, pitching for the Giants. It's a miracle."

"Hey, sport, he's only flesh and blood."

"He's marvelous, Eli, just marvelous, and I have to let him know that I know it."

"With this ring?"

"Yes!"

Eli looked at me, and at the drawing, and at me once more. "Then we'll do it," he said.

"We will?"

"Yes, on one condition. No, seriously, sport. This"—he tapped the drawing—"is the first design you've done since we left home. You've a lot more to do, sport, that's what this trip is for. I wasn't kidding Uncle Sid when I told him you needed to learn the market. This is no vacation, it's your living. Mine, too, and the whole family's. Thank God you're good at design! If you had to sell we might all starve. But good as you are, Jackie, I know that a design like this comes easy to you. You've got to start on the hard stuff, like trying to guess what ten thousand people are going to want to buy for grandma next Christmas. That's what pays the rent, and that's what you're out here to learn. You can't sit in a corner waiting for the presentation to end so we can get out to the ballpark. You have to talk to these people, do sketches and show them around, listen to opinions. Now you get started on that, and we'll see to this ring when we get home. Is it a deal?"

"How are you going to get the ring to Mathewson?"

"Me? This is your pot to stir, sport."

"Oh, no, I couldn't. I really couldn't."

Eli sighed. "All right, I'll take care of that too. But you've got to get your ginger going. We've two more days here in Chicago and then two in Detroit, two in Cleveland, one in Buffalo, and one in Albany. When we get back to New York I want you to have a book full of sketches to show Uncle Sid so he won't bite my head off when I hand him the bills for the trip. You won't have to do it alone, sport. I'll help you, and so will the clients; they want you to give them what to sell, and they're not bad folks. You'll find them a lot easier to talk to than a couple of son-of-a-bitch ballplayers. Or even Christy Mathewson."

Bucknell College, Pennsylvania
October 10, 1901

My dear Mr. Kapp,

The ring honors me greatly, and I could not appreciate it more. I have already expressed this to your brother, a very entertaining fellow, and I hope that you will extend my best wishes to him and to the craftsman who brought your design to its beautiful realization. I assure you that it will ever serve as a treasured reminder of the occasion and the efforts of my teammates which made it possible, and also of your artistry, which in its field certainly surpasses my own in mine.

I pray that you will overcome what your brother describes as an innate modesty so that we may meet and I may express to you what this poor note can only attempt.

I am, with great courtesy,

Yours,
Christy Mathewson

The letter told me what Eli had not: the excuse he made for me when in the final week of the season he delivered the ring. Mathewson had pitched his twentieth victory of the year and the team's fifty-second and last, and I stayed in the grandstand, hardly more empty after the game than through its progress, while Eli visited the clubhouse in right field. I feared for him, imagining that Warner and Strang had alerted the entire club to the spectre of ingratiating Israelite gem-peddlers; I had a vision of Eli, beaten and spat upon, rolling down the clubhouse steps like a battered barroom drunk. No such episode occurred. Eli reported that if Warner and Strang had noticed him at all they gave no sign of it. As for Mathewson, he appeared fascinated that his achievement had moved a spectator to create such a souvenir. He neither declined the ring nor offered to pay for it. In short, I thought, he accepted it in the spirit of its giving.

I did not respond to his letter. From the time we returned to New York I felt increasingly embarrassed at the whole progress of events. If Eli had quietly let the matter drop I'd have made no protest, but he carried it through and I had to fulfill the condition he'd laid down. That winter, with Edith off to a Swiss finishing school and Eli out of town more often than not, I spent every hour at design. I used the gymnasium not at all; my body went soft and white, and my clothes grew tight around my waist and buttocks. I visited museums, often with my youngest brother Arthur in tow; I was expanding my vocabulary of design, and my mother had artistic aspirations for her seventh and last son. In December we saw a display of Greek statuary at the Brooklyn Museum. I especially admired the athletes, gods in marble.

The winter baseball news bore aspects of obituary, so frequent were the reports of departure and bewailing. Players, managers, even owners jumped or threatened to jump to the new league or back again. Continuity died; bidding wars raged; the most nondescript veterans gained instant paper fortunes as salaries trebled; vituperation became the language of the day. Nor did the opening of play that spring stem the chaos. The American League now challenged the National in half its cities, and a player might enjoy a successful homestand in the older association, then skip across town to try his luck in the newer for greater pay. Managerial discipline vanished and the concepts of sacrifice and team play were abandoned.

The interleague traffic gave me one reason to applaud: Warner and Strang jumped, the catcher to Boston, the infielder to Chicago. The Giants began the season with a dozen new players, none of distinction. Owner Freedman, standing aloof from the year's mad auction, gained neither a pristine reputation (what Tammany man could aspire to that?) nor a competent team on the field. There was yet another manager, Horace Fogel, a relic of the old Indianapolis club of the 'Eighties but, to Freedman's satisfaction, a registered Democrat. He was fired in June, succeeded by second baseman Heinie Smith. Smith gained credit for recognizing talent and obloquy for misusing it. It being obvious that Mathewson was the best of his athletes, Smith determined to play him in the field and at bat between pitching assignments, and we few—very few—Giant supporters saw him now in the outfield, now at first base. Once in Smith's reign I came early to the park and observed Mathewson working out at shortstop under the manager's tutelage; the pitcher seemed as amused by the exercise as I was horrified, and I felt then that I would rather he jump to a team that would use him properly than witness this grotesquery. His record suffered; how could it not? He lost his fine edge, allowing a base on balls here, a base hit there, a single run that often proved decisive. Nor did the Giants support him, even to the miserly extent of the previous season.

There was loud complaint from every corner save one. While athletes pouted and departed, while fans cursed players and management with equal vigor, while editorials decried the times, from Mathewson one saw only honest effort on the field, and heard nothing. If I attended a game with Eli I'd take the East River train home while he visited the clubhouse; if I went alone I'd linger after the game as the Giants, in twos and threes, drifted out. The purpose of the vigil was a glance at Mathewson's right hand; the reward, a sight of the ruby ring.

With Eli on the western swing that summer I discovered the depths of national disenchantment with the professional establishment. Many of our clients were frankly disgusted with big-league ball; the pastoral game where local heroes bore the honor of the city against the invading foe had transformed into a disquieting brawl among mercenaries. If this saddened the affectionate fan, it absolutely threatened Eli and, by extension, the emerging prosperity of the family. In his integration of baseball and business "Kappy" was unique and famous among his clientele. Now his invitations to the park met refusal, polite but definite, and he worried about its effect on sales. In Cleveland we found fewer than a hundred spectators at an American League contest, despite Napoleon Lajoie's presence on the home roster—or perhaps because of it. It was Nap's third shift in as many seasons. He'd batted an astounding .426 for Connie Mack in '01, and the Philadelphia Nationals, his original club, suffered greatly at the gate. In the spring they went to court to regain Lajoie's exclusive services; on Opening Day a sheriff marched to home plate in the first inning to slap Lajoie with an injunction forbidding him to play in the state of Pennsylvania for any club other than the Nationals. Nap tossed the papers into the air, bid the pitcher throw, and slapped a two-base hit while the sheriff stood beside him in the box. But it was his last hit for Mack's Americans. After the game he was smuggled to Cleveland, and there he played out the year, unable to cross the Pennsylvania line with his club. Little wonder that the ball fans in Cleveland—Napoleon's Elba—considered him a temporary attraction and paid him small regard.

Our path did not cross with the Giants that summer. The anniversary of the no-hit game found us in Buffalo while the ball club was in Cincinnati, losing yet again and falling further into the National League cellar. On Wednesday we were in Albany, and the next evening we boarded the night train south toward home. Eli added up his accounts while I sketched busily.

"Hey, sport, why not get your head out of that pad for a bit?"

"And do what?" I asked.

"All work and no fun——"

"I hadn't realized that fun was our purpose."

Eli looked at me. There was no merriment in his eyes; there'd been little of it in the month we'd been on the road. I turned a page and quickly roughed out a sketch.

"Here," I said. "The New York Giants club ring of Nineteen and Two." It was the ruby ring, but in parody: the gemstone was a black pearl, the baguettes chips of coal, the band etching rows of zeroes. For engraving there was this slogan: *Ad Mediocritas.* "How much would it cost to make up?"

"Gold band?"

"Tin."

"I don't know. Black pearls run high."

"Make it an onyx, a faulted onyx. I saw a couple on the floor at the shop. What would it come to?"

"About six bits. Times twenty-five, that's . . . eighteen seventy-five, a ring for everyone on the club."

"Not everyone," I said.

"No?"

"First base. Can you imagine, playing Mathewson at first base?"

"He's a good first baseman."

"He's a great pitcher."

"And a grand guy to know, sport. Why don't you come with me and let me introduce you, next game we see?"

"No, thanks."

"You're a strange one, sport."

"There'll be a time, but not now. Certainly not now."

"Now's the best time, sport. You don't want to be a fair-weather friend, do you?"

"I don't want to be his friend at all."

"A secret admirer?"

"No secret to it."

"What, then? A worshipper from afar?"

"Isn't that the proper distance for worship? You don't crawl into the ark to worship *torah.*"

"No, but you touch it when it's walked through the temple."

"I've never liked that, people climbing all over one another while the scrolls are paraded through the aisles. It's not dignified."

"It's emotion, sport, like cheering at a ballgame."

"I suppose." I shaded the drawing on my pad. "Eli, you don't really think I worship Mathewson, do you? In the religious sense? That's heresy."

"Yes. I mean it would be heresy, wouldn't it?"

I considered the thought. "A very American heresy, Mister Kapp. Mister E. Kapp."

Eli laughed and affected an accent. "I'm a *yenkee duddle* boy."

"And now the batteries: pitching, Jakie the Jew; catching, Eli the Hebe."

"Strike *ein!*"

"Strike *zwei!*"

"Strike *drei, unt raus!*" It was good to laugh together.

When the train stopped at Hudson Eli asked the porter to fetch the late newspapers from the city. The Negro brought back a copy of the *World,* Pulitzer's yellow sheet; by old practice Eli ripped out the mercantile listings for his own study and handed me the rest. I opened to the sports pages.

"Oh, my God!"

Eli looked up. "What's the matter, sport?"

"Freedman's fired Smith——"

"So, *nu?* Another month, another manager."

"And he's hired Muggsy McGraw."

Eli dropped the paper. "McGraw! No!"

"And half the Baltimore team is coming with him."

"Who?"

"McGann, Bresnahan, McGinnity——I wonder if he's bringing anyone who isn't Irish."

"Robinson?"

"No, Robby's staying to manage the Baltimore club. But he's got nothing to manage. Listen to this: 'Baltimore Forfeits. St. Louis Club Takes Field Unopposed. Orioles Cannot Raise Nine Players.'" I folded the sports sheet. "The old Orioles, gone. First Hanlon takes half of them to Brooklyn and they drop out of the National League, and now McGraw takes the rest to New York and they can't field a team in the American. It's unbelievable. Oh, Eli, remember? Remember Keeler, with his hands way up here, peeking over his elbow and just—pop!—pecking at the ball . . ."

"Remember Brouthers? First time I saw Big Dan I took a look at those arms and said, 'That's no ballplayer, that's a blacksmith!' "

"Not a real Oriole, though. You think of the little fellows, the bunts and steals and the hit-and-run."

"Jennings."

"Yes, and Reitz and Kelly."

"And McGraw. McGraw! Can you believe it, McGraw managing the Giants!"

"Managing Mathewson. How do you feature that? The gentleman and the ruffian in the same uniform. I can't see it, Eli. Mathewson will jump."

"Don't bet on it, sport."

"I can't see it any other way."

"You don't know Mathewson, sport. He loves to win, and McGraw's a winner."

"He'd rather win fair and square than McGraw's way."

"Don't be silly, Jackie. I know the man, I've watched him play poker."

"Poker?"

"Matty's a gambler, sport," said Eli, retrieving the paper from the floor. "He spends his Saturday nights playing poker in a room at the Victoria Hotel."

I considered the information and its source. "And takes you down to your garters, too, I wouldn't be surprised," I said.

Eli looked at me and said nothing. I broke off the stare to read more news. The next sound in the compartment was the porter's knock. We undressed separately; not until we had lain in the dark for some minutes did Eli break the silence.

"We're going up there Saturday, aren't we, sport?"

"I wouldn't miss it," I said. "Not for a contract with the Giants."

Rarely since the Temple Cup series eight years before had such a crowd besieged the Polo Grounds. Thousands milled about the grandstand gate, ringed the outfield, spilled along the foul lines and mingled behind home plate. They shared my own state of excited curiosity to see John McGraw in a New York uniform, for as he wore those colors we all had to learn to love him, or to hate the Giants; touching McGraw, nothing else was possible. He had strutted across a decade of Oriole triumphs like a bantam cock, insufferable in victory, petulant in defeat, courting the hatred of every baseball fan outside Baltimore, the most thoroughly despised man in the league and thriving upon the fact. No one had contributed more to Oriole victory, none had so embodied the Baltimore style of unremitting pressure upon the opposition. Together those marvelous players had achieved a cooperative genius that first dominated the league and then transformed it into a pack of imitators. But to McGraw the game never stopped. His combative fire raged as fiercely off the field as on as he bad-mouthed opposing teams, insulted entire localities, and fought, constantly fought, with hometown heroes, with umpires, with fans, owners, newspapermen—fights that brought the multitude to a frenzy. His name called forth images of police escorts, rotten eggs, scalding oaths, and the frightening fury of a baited mob.

We caught first sight of the crowd when the train took the hard right on Harlem Heights and began the slow descent to the river; we knew immediately that it would be futile to seek a grandstand entry. Grinning at the memory of days long past, we headed for Coogan's Bluff. Along that high ridge west of the grandstand a thin line of men had formed, and at its northern extension, which offered the least obstructed view of play, they were shoulder to shoulder. We climbed the slope of the bluff and ran; I outpaced my brother by fifty yards and pushed in among the eager cluster on the ridge. Eli, panting for breath, found me, and as when we were children we edged a short way down the steep incline and sat precariously on a huge boulder that protruded from the cliffside like the bow of a buried ship. We were not long alone; the rock had safe room for six or seven others, but twice that number tried to squeeze in. Some slipped and needed rescue. Common danger and common concern sorted things out, and we settled, uncomfortable but reasonably secure, a hundred feet above and two hundred yards from home plate.

A great shout arose, and the mob below rushed for the clubhouse as the Giants emerged. A wedge of police led the players on their progress as men threw their hats in the air and waved their arms. The arrowhead cut its way down the foul line toward home plate, where the blue-suited umpires and the Philadelphia manager waited. The wedge separated, and McGraw strode forward to join the meeting, a minuscule white mouse emerging from a maze, a ludicrous dot in the scene. From our remove he was featureless; memory more than sight delineated the wiry body, the thin arms, the triangular face, and the jug ears that stuck out from under his cap, foolish and duncelike. There was little in the man's physical makeup to admire save the savage energy that seemed to overbear his stature and escape in quick, vigorous movements. As the ovation rose he lifted his cap and waved it at the grandstand. Some of the men who shared our perch joined the applause, and I was surprised that they could accept McGraw as their own so swiftly. So was a large derbied man among them.

"Sit down, sit down. He's done nothing yet for your cheers," he said.

"Ah, but he'll bring better days, Johnnie!"

"He'll have to go a distance to make up for the bad ones he's brought upon us, Davie," the man answered. "I see 'New York' across the man's chest, and I feel as if an Orangeman's been made pope."

In the narrow space the police had cleared in front of the Giant dugout two players began to throw a ball back and forth; one was diminutive, the other big and broad. I thought at first that the latter was Mathewson, but when he bent into his motion I knew at once that it was not.

"He's starting McGinnity!" I said to Eli.

"Yes, and that little fellow must be Bresnahan. A Baltimore battery."

"But Mathewson's in turn!"

"The manager's brought his own men, sport."

"That's the best thing about this business," said the man with the derby. "We've got a real pitcher at last."

"What's wrong with Mathewson?" I said.

"Mathewson? Just a trick pitcher."

"A trick pitcher!"

"Yes, and they've figured him out, haven't they! He's lost more than he's won this year. Sure, he was all right his first time through the league, but they've wised up to him by now."

"You know nothing about pitching."

"What's that you say, sonny?"

"Take it easy on my brother," said Eli. "He's stuck on college men, reading books as he does."

"Well, give me a good Irish hardballer any day."

"McGinnity is that," said Eli.

"The man's an ignoramus," I muttered.

"And we're in Mick heaven, so keep your mouth shut."

Viewed from a distance, McGinnity's strengths were certainly more apparent than Mathewson's. The barrel-chested black Irishman was as unsubtle as the steam locomotives that brought up more and more eager spectators until there seemed as many in the outfield as in the packed grandstand. A steady clamor rose from the flats and burst into thunder when the Giants ran out onto the diamond to take their positions.

"He's at shortstop! McGraw's at shortstop!"

"Guts," said the derbied man. "Aye, he's got guts."

"He's no shortstop."

"He was, long ago."

"Not a good one. Baltimore didn't win 'til they moved him to third base."

"It's command," said derbied Johnnie. "He's taking charge of the club, and shortstop's the place for it." I warmed to him a bit, for he knew something about baseball; shortstops steer the defense. We strained to identify the other starting players. The Baltimore contingent included big Dan McGann at first and Steve Brodie in center field, along with McGraw and the McGinnity-Bresnahan combination. McGinnity was in form, predictable and overpowering; each pitch burst into Bresnahan's mitt like a gunshot, and as Philadelphia batters swung and missed the Irish on the rock cheered. "Now you're seeing the real thing," Johnnie assured me, and Eli gripped my wrist to ensure my silence. My brother joined the applause as the last Phil went down in the first inning. Jones drew a walk to start the Giant half, and here was McGraw, at bat in our home uniform.

I looked at the faces on the rock and guessed at their thoughts. Were they remembering '94, when at the moment of our victory McGraw dismissed the Temple Cup series as mere exhibition play? Or the game the following year when, attempting to hold a runner at third, he'd torn off the man's belt? Or his charge into the grandstand with Kelly and Doyle, bats waving, spikes scraping the astonished spectators?

I watched Jones lead off first and whispered "Oriole baseball" to Eli, who nodded and offered two bits that McGraw would bunt the runner to second. I was surprised; these were no clients to jolly along. There was a taker, and Johnnie offered his derby to hold the stakes. A moment later, the sacrifice accomplished, Johnnie emptied the coins into Eli's hand. He poured them back, saying, "Four bits that he scores." When no one answered he simply reversed the offer: "All right, four bits says he doesn't!" Now two men accepted the bet; Brodie's scoring double cost Eli the pot. He paid cheerfully, for the Giants led by a run.

Philadelphia caught up with McGinnity in the third. The pitcher's fastball was a fierce weapon, but double-edged; it required an expansive wind-up, his arms swinging like huge pendula, his body pumping twice. With men on base he was forced to abridge the motion and couldn't generate the same power. I said as much, and Eli's quarters clinked in the derby against the offerings of the loyal Irish. The Philly attack continued: a hit, a walk, a scoring fly ball. McGraw interrupted the rally with a senseless rage at the umpire, insisting that the runner had left base before the catch, and although none of our company had watched the man they joined in the argument:

"Why, he's right!"

"Sure he's right! Didn't you see it?"

"I was watching plain. He left too soon."

"Tell him, Mac. Give him the devil!"

"Robber! Bloody landlord! You're blind!"

It went on too long, and couldn't have helped McGinnity, but McGraw argued to the limit; I was sure that umpire Emslie would have tossed him out if he thought there was a chance, having done it, of leaving the grounds alive. Finally McGraw took up his position and play resumed. There was a sharp ground ball to shortstop, and McGraw took a gamble, throwing to third to cut down the leading runner. It would have succeeded if third baseman Lauder had expected the toss, but he'd slept through three months of the season without extending himself, and he muffed the tag. The glare from McGraw must have burned poor Bill Lauder's skin.

Eli and I shared a whispered exchange while the Irishmen

waited impatiently. I weighed what I knew of McGinnity against what I hoped for the Giants and advised Eli to go against the pitcher. He announced his bet; the coins dropped into the hat. A strike, a ball, another ball, and then a solid base hit that put Philadelphia ahead. Eli merrily pocketed his winnings while Hughie Jennings came to the plate. We savored the irony: Jennings had replaced McGraw as Baltimore's shortstop years before.

"Well, Johnnie, which Irishman do you like here, Jennings or McGinnity?" said Eli.

The man turned to me. "Where do you put your money, son?" he asked.

Eli laughed. "I'm your man, Johnnie. Jackie's just along for the fun."

"Declare yourself, sonny," the man said, holding the derby before me. I took out my billfold and dropped it, unopened, into the hat. "McGinnity's beaten," I said.

"You're on, sonny, for whatever that wallet holds."

The first pitch to Jennings was pure McGraw; McGinnity was merely the instrument. It flattened the batter, passing his head by a fraction; Bresnahan barely snagged it with the web of his glove. "Sweet Jesus!" said a voice on the rock, and another man crossed himself.

"Look at Jennings! He's staring at McGraw!"

"Mac's laughing. Look at him!" I watched McGraw; his mitt and bare hand were cupped around his mouth, and I wondered if he was shouting encouragement to McGinnity or threats at Jennings.

"If I were Hughie, I wouldn't dig in for this pitch either," said a voice behind us. The anticipation was intense as McGinnity set and threw. Jennings swung; the ball racked off his bat toward shortstop; McGraw leaped, but the ball cleared his glove and bounded into the outfield. Another run scored, and my bet was won.

"Get the bad news, Johnnie!" Eli cried amid the uproar, and the man reached into the derby, extracted the billfold and opened it.

"Hah! Empty, is it!" I'd palmed my money while the wallet was still in my pocket. He reached into his jacket and extracted a brown whiskey bottle. "Well, I've got something that's full enough. Have your fill, Jocko me boy. You've a nerve, I'll say that."

"I backed the better Irishman," I said. "To your health, Johnnie."

It was awful whiskey; I winced and closed my eyes as it burned my insides. When I looked up another Philadelphia run was crossing the plate. I passed the bottle back to Johnnie and it went around the rock, taking the sting off the three-run Philadelphia lead. This was a daring compensation to attempt, given our position, and as the innings passed it created an adventurous suspense that quite overwhelmed the diminishing drama below. McGinnity regained his touch and the Phils scored no more, but the visitors' pitcher held off the Giants with equal effort until the seventh. The Giants scored then, costing Eli a bit of his winnings, but the lead was still two runs when McGraw batted with two out and none on in the eighth.

A short ninety minutes had not erased the antipathy of a decade. He was still McGraw, and there needed more than a home uniform to make him ours. For myself, that he had passed over Mathewson for McGinnity weighed heavily against him. True, he'd never seen the collegian pitch; true also, he must establish command of the club, as derbied Johnnie had understood. Yet I had to think that his advent meant Mathewson's departure; these opposites could not attract.

McGraw was a hitter, undeniably; in his best years he'd batted fourth for the Orioles, and that, more than any of his antics, was how he'd hurt us most. Now he was a Giant hitter, and then, with a base on balls, a Giant baserunner; after Brodie's double he scored a Giant run, and Philadelphia's lead was down to one. Once more the hollow echoed with shouts and urgings. Lauder walked, the bases were crowded, and Giant hopes rested on Bresnahan.

"Anyone betting on the new man?" said Johnnie, holding out the derby. A few coins went into the pot, and Eli matched them. Bresnahan went out weakly, and my brother took the final pot of the afternoon. The Giants went swiftly in the ninth, and another bottle was passed to wake the defeat.

"They're a better club, no doubt of it," said Johnnie. "They made it close."

"But no cigar."

"I say them welcome, nevertheless. You're right, Davie, there's better days to come."

"They didn't give up."

"They fought back bravely."

"Kept their spirit, didn't they? After all, they lost by a mere run."

"And that a gift from the bloody umpire."

"They'll be out of the cellar before long, I venture," Johnnie said. "McGraw will take them out, sure."

"If he uses Mathewson well," I said.

"Little I care about Mathewson, Jocko. There's no faith in a Scotsman." Grinning, he offered me a final pinch from the bottle. I emptied it and tossed it down the bluffside. "Johnnie, you're a fine fellow," I said, "but you know as much about pitching as you do about good whiskey."

"Well, Jocko, you bring the bottle next time and see what you can teach me about both. Now there are two ways off this rock, and I'll need help on the upward path."

There was a great crush to gain the stairway to the elevated train platform. Eli and I waited near the clubhouse while the crowd thinned out; the drift we overheard matched the consensus on the rock, much satisfaction with the club's showing, little distress over the one-run loss. A train departed, the gabble diminished, and in the quiet we discerned a single voice that issued from the windows of the clubhouse above us. Angry, vibrant, profuse in its profanity, the voice caught all ears and silenced the listeners until it was the only sound in the air. It was a grating, pitiless voice, boring like a drill, scolding the ballplayers, accusing and threatening them, and telling them, telling us, that there was one in that clubhouse who was not content with defeat, no matter how near.

"God of light!" Eli whispered. "You'd think they just lost the Temple Cup!"

I imagined Mathewson in the clubhouse, blameless, buffeted by the vulgarity and embarrassed for his teammates; I half expected to hear his voice against McGraw's, insisting on fairness and moderation. I turned for the train and realized that Eli had not moved.

"Come on," I said.

"No, I'm staying. I want to welcome the man to New York."

"And say goodbye to Mathewson?"

"You don't understand the matter, sport. Winning forgives all."

I left Eli standing there and climbed the long steps to the station platform. A train puffed in, its black smoke rising against the dusky sky, and I looked back at the faraway clubhouse and saw my brother stepping gingerly to its great iron door.

October 14, 1905—at New York

										R	H	E
PHILADELPHIA	0	0	0	0	0	0	0	0	0	0	6	0
NEW YORK	0	0	0	0	1	0	0	1	X	2	5	2

BATTERIES: Philadelphia, Bender & Powers;
New York, Mathewson & Bresnahan.
Winning pitcher: Mathewson. *Losing pitcher*: Bender.

THE WORLD'S CHAMPIONSHIP SERIES

	W	L	PCT.
New York (N.L.)	4	1	.800
Philadelphia (A.L.)	1	4	.200

New York wins best-of-seven series, four games to one.

TWO

I FEEL slightly ridiculous!"

Eli braked the car and strained to hear my confession over the evening clatter of Union Square. I repeated it with a shout:

"I feel slightly ridiculous!"

"That's all right! You look totally ridiculous!"

"So do you!"

So did we both. Still, I felt that Eli presented the more comic picture, balancing a high top hat on his head with one hand while he gripped the steering rudder with the other. My own topper rested neatly on the lap of the white riding coat that protected my evening clothes. The clothes were rented, but the Daimler automobile was not. Eli's winnings had paid for it. From the first day of the 1904 season to the pennant-clinching victory in mid-September he'd laid ten dollars down on every Giant game, twice that amount when Mathewson or McGinnity pitched; his net gain topped a thousand dollars, and after the purchase of the car there was yet enough left to spend on a pair of tickets to the team's victory celebration at the Erlanger Theatre that evening.

We'd watched many of the games from our box seats in the second deck of the Polo Grounds—seats that hadn't existed two years before. As McGraw built the club on the field, a new ownership embarked on construction to accommodate crowds beyond the imagination of the past, thirty thousand, thirty-five thousand. All of New York seemed to be building; beneath us as we proceeded up Broadway ran the subway tunnel that would open in weeks, inaugurating a rapid transit system that linked the Battery to the Bronx; above us to the west stretched the elevated line I planned to ride between work and my new digs off Ninth Avenue in Chelsea. The Erlanger Theatre itself, at Forty-Fifth Street, was an advertisement that the theater district no longer ended at Forty-Second.

Traffic slowed again at Madison Square. We idled in front of the ornate facade of the Garden. I reached into my waistcoat for my watch.

"Don't worry, sport, we'll be there in plenty of time."

"I don't want to miss a moment of it."

"Neither do I, sport. I need an answer."

"Oh, that. It'll be Mathewson, no doubt."

"If it were up to you, McGinnity wouldn't pitch at all!"

"Of course he would," I said, "but in the first game of a World's Series you have to go with your best."

"Who will Boston throw?"

"Dineen."

"Not Cy Young?"

"Dineen did the trick last year."

"Wasn't that a whoop? I never took such a bath." Conversation stopped as we began to chug northward, and picked up again when a westbound trolley blocked our progress at Thirty-Fourth Street. "It was your fault, sport," said Eli. "You told me the Americans didn't stand a chance against Pittsburgh."

"I'd have been right, if Wagner could have hit a lick."

"Would have, could have. There'll be a pile of money floating on this Series, sport, and I want to be on the right side this time."

"You'd bet against the Giants?"

"Mum's the word. I wouldn't ruffle your dainty feelings so close to your wedding day."

"Just don't lose the car before the wedding, Eli. I want it waiting when we come down the steps."

"If things break our way in the Series, sport, you'll have one of your own."

I grasped the top of the windshield and stood to look ahead. "Goodness, Eli, what traffic! Half the autos in Manhattan must be heading to the Erlanger!"

Not only automobiles but hansoms and hacks and sleek brass-trimmed broughams jammed the approaches to Longacre Square and the new Times Tower. We turned west at Fortieth Street and searched for a livery stable that would accept care of the Daimler.

We were halfway through Hell's Kitchen before we found one. We folded our riding coats into the toolbox compartment, and I chafed while Eli displayed the fine points of the auto to the wide-eyed stablehands. Finally we took to our feet, walking north to Forty-Second Street and then east along the grand theatrical promenade. Some of the houses were shuttered—the Amsterdam, the Victory—so that their star attractions could entertain at the Erlanger. It was a night for New York's grandees: fur-trimmed coats and opera capes whirled about under the Erlanger marquee, which announced the event in bold letters:

NEW YORK GIANTS
NATIONAL LEAGUE CHAMPIONS
1904
GALA VICTORY CELEBRATION

How familiar the public faces around and inside the theater, how engaging to see them in lively animation! Here were politicians out of cartoons and newspaper etchings, theater folk in civilian costume, mercantile princes whose portraits stared down from the walls of their banks and emporia. Fascinating to watch Tod Sloan and George M. Cohan in a handshake—Sloan, the white jockey who'd won the acclaim of the haughty British, and Cohan, who'd turned the tale into the grand musical success of the season, *Little Johnny Jones!* Wonderful to speculate what imperious instruction Tammany Boss Croker whispered into Mayor McClellan's obedient ear! Thrilling to encounter at every turn the men of the moment, the athletes, bronzed and lined by the summer sun yet far more youthful, face to face, than they appeared from the decks of the grandstand. Determined to touch every arm and hear every aside, Eli plunged into the orchestra with a will. I edged toward the balcony stairway, muttering pardons. A hand gripped my arm; I turned and found myself face to face with Jack Warner.

He'd come back to the Giants in '03 with the peace settlement between the leagues and the establishment of the National Commission, a .200 hitter and a faint echo of the days before McGraw. He owed his place on the club to his local popularity in the small section of upper Manhattan that was his home, but I didn't doubt he'd be gone soon; the champions had little need to court a following by carrying a home-town hack. He held my arm. "Aren't you Kappy's brother? The pitcher? Sure you are! Say, remember that time we had dinner together in Chicago?"

"I remember it," I said.

"Sure, Kappy and I have had some laughs over that! He's a whale of a fellow, your brother. Good poker player, too. Why don't you come with him to the Victoria some time?"

"Poker's not my game."

"You still have that bee in your bonnet about Matty? That sure gave me and Dixie a howl!"

"I guess it did."

"Great night, hey? Where's Kappy?"

"In there, somewhere."

"That's Kappy all over, ain't it? I guess he's told you about the job."

"Job? What job?"

"You know, me showing your stuff to the boys on the club. I hope your uncle goes for the idea. I can really help out."

"I don't know about it at all," I said. "I just design the line."

"And you're the best, kid. Say, that ring of Matty's, it's swell."

"I liked it."

"So do I!" He put his hand on my shoulder. "You know, kid, I always thought that Dixie was a little rough on you. I told him so, many times."

"I guess he was." I shrugged, and Warner released me. "Good to see you again," he said. "Your name is . . ."

"Kapp," I said.

"No, I mean——" He stopped. "Nice seeing you."

"Sure."

I gained the stairway and pushed up to the balcony, checked my ticket, and found our seats in the last row at the left wall. I put my top hat on the seat and climbed down to the railing. The orchestra aisles were packed with clusters of the fancy, laughing, exchanging greetings, pointing at the greater celebrities.

There needed no spotlight to magnify Mathewson. He stood at the foot of the steps to the stage, elegant in formal attire, shielded from the crush by a bank of respect. John Brush, the club's new president, stood beside him, and an underling brought forward each new set of well-wishers as the last turned away. I saw Eli working his way to the fore, and as he looked about he found me and waved; then, pointing at Mathewson, he beckoned me to join him. I shook my head; I thought that point had long been settled. Eli dismissed me with a gesture and edged forward, standing tiptoe to glimpse the pitcher. Finally his moment came, and he greeted Mathewson, then Brush. The three passed words, Eli turned, and with his arm about Mathewson's shoulder he pointed up at me.

Mathewson smiled at me, nodded, and watched as I thought what to do. Certainly I owed him a salute. The mastery he'd displayed three years before in St. Louis had become nearly standard in the 1904 season, his league-leading numbers a mere shadow of his craft, not its measure. In one month I'd seen him pitch four times without issuing a single walk. When the Giants furnished him a comfortable lead he would coast, but if a potential danger arose he was instantly transformed, and the threat was stifled. He was at his best against the strongest competition in the closest contests. In Mathewson and McGinnity, the pure pitcher and the primitive thrower, the Giants had a tandem unequaled in thirty years of National League play, and no club had ever won as many games.

I took my hands from the railing and began to applaud very quietly; my homage, I hoped, would be a private one. He bowed slightly in acknowledgment, and when he straightened I returned his smile. Eli grabbed Mathewson's wrist and held it high, pointing at the ring on the pitcher's right hand. The athlete forced his arm down and remonstrated. I winced and turned away, humiliated for myself and for Mathewson. I retreated to my seat in the dark corner of the balcony and waited for the great show to commence.

As the chandeliers dimmed a spotlight played over the orchestra pit, and the maestro gave the downbeat for "Columbia, The Gem of the Ocean." Everyone stood to sing along. The curtain rose to reveal a painted backdrop of the Polo Grounds, and the cheers matched any I'd heard at the park, rolling on as a color guard marched down the orchestra aisle and onto the stage. A trumpet soloist played a fanfare, and then, with a leap, George M. Cohan emerged from the wings. He frolicked amidst the guardsmen, stiff at attention, and then he whirled forward, spread his arms, and broke into an anthem about Broadway and Herald Square. Cohan's lively magic provided a fitting introduction for jockey Sloan, the evening's master of ceremonies. The speeches began, all boasts and brags: praise to the team, to the city, to the great game of baseball; encomiums for the peace between the leagues, with references to peace between nations and among peoples; then, in a swift and ironic about-face, dire predications of the awful fate that awaited the Boston Americans in the World's Series to come.

Eli slipped in beside me, and I ignored him. Angry at the Mathewson episode, confused at the exchange with Warner, I wondered what other dealings he had of which I knew nothing. I found myself stealing sidelong glances at his face while he watched the Giant players trot up the aisle to receive diamond clips: Warner, Bowerman, Taylor, survivors of the grim cellar seasons; McGann, Bresnahan, Dunn, Browne, the Oriole veterans; Gilbert, Dahlen, and Devlin, the bright young infielders; and the right-handed jewels of the roster, McGinnity, thirty-three victories, and Mathewson, thirty-four.

As I stood and cheered, Eli struck me on the back with emphatic force. "Oh, sport, why didn't I think of it!" he said.

"Think of what?"

"Rings! Championship rings, one for everybody on the club! God of light, why didn't I think of it!"

"Next year, Eli."

"No, no, this year, for the World's Series!"

"Can we afford it?"

"We'll sell them, sport. At cost, but we'll sell them to the club. I'll see Brush about it as soon as this is over. Meet me at the car!" He was off down the aisle with a rush; I half expected him to burst into the spotlight when it played upon the first box, illuminating the club's president. Then it danced back to Sloan. I watched the athletes behind him shift about like kindergarten boys while he introduced Harry Stevens, the Polo Grounds concessionaire, for the evening's final presentation; they snapped to respectful attention when Stevens stretched his arms skyward and cried out the name of John McGraw.

He pranced onto the stage like a show pony, head high, jaw thrust forward, eyes shining in the arc lights. Never was there such an ovation: the crowd was on its feet, some standing on their seats, and high hats tumbled in the air. The little man turned to dismiss the ballplayers and joined in the applause as they marched down the steps. The clapping became rhythmic and more rapid; finally it exploded into a primitive, encompassing roar of triumph. The fourth time that Stevens held up his hands the clamor began to subside; then it grew and lessened by stages until his words could be understood. His presentation to the manager—a set of diamond cuff links—unleashed yet another outburst. Finally McGraw had the stage to himself. He included so many in his salutation that I felt positively ignored that my own name went unmentioned; at last he took me in with the phrase "baseball fans of New York."

He thanked Stevens for the gift, then began to speak of his early days in upstate New York, where the city seemed as distant and as glittering as Babylon. He acknowledged his early instructors and teammates; when he listed his Baltimore cohorts the audience booed the names, but with good nature and laughter. He didn't overlook the newspapermen, nor their editors nor publishers; he found kind words for Freedman and better ones for Brush. He revealed a political tie to Tammany with a joke about Roosevelt and an endorsement of Judge Parker for President. When he came to his players he had only perfunctory praise for the Giant veterans he'd found on his arrival; Warner was among these, and I thought of the catcher's pleading in the lobby. He warmed as he talked of the Baltimore imports, especially Bresnahan, who'd filled in wherever needed and ignored persistent injury to do it. He joked that the youngsters Dahlen and Devlin had forced him to retire to the bench, but warned with a fierce smile that he was ready to take up the glove once more if need be.

Then the pitchers. "Joe McGinnity," he said, and the crowd was up and cheering.

"Joe McGinnity," he repeated. " 'Iron Man,' the newspapers call him, and that fits a man who has three times pitched both ends of a double-header. He is a marvelous pitching machine. All I have to do is point him toward home plate and say 'throw!' and he will throw and throw until he drops. But I'll tell you a secret about Indian Joe and those double-headers. He can throw, but he can't count!" Everyone looked at McGinnity in his aisle seat; as he smiled, it gave us leave to do the same.

"So in the third inning I told him it was the second," McGraw continued, "and in the fifth I told him it was the third. In the ninth I told him it was the fifth, and when the second game began I told him it was the sixth inning of the first. And I would say that Joe didn't know he'd pitched a double-header until he read it the next day in the newspapers—except that Joe can't read the newspapers!"

McGinnity roared, and so did the whole house.

"So I have a pitcher who can't read the newspapers," McGraw said, timing his remark to the dying laughter with the polish of a vaudevillian, "but I have another who not only reads them—he corrects the grammar and the spelling!" Laughter welled up and dissolved into shouts of "Matty! Matty!" The hero stood and waved; the ovation, I was happy to note, at least matched that for McGinnity. Only McGraw had received a greater one.

"I knew this young man only by reputation when I came here," McGraw said. "I was told that the New York roster boasted a collegian who was capable of almost anything on a ball field—he pitched, he played the infield and outfield with equal skill, he batted as well as any regular. I was eager to see this paragon, as you might imagine. I even hoped I could save Mister Freedman some money by dismissing the entire roster and allowing the young man to play single-handed. But after all, I'd brought some excellent athletes with me, and found some very good ones here, and it seemed a waste not to employ them, so I apologized to the young fellow for my seeming disrespect and asked if he might share the diamond with some seven or eight others. He is a gracious chap, and he granted me the favor."

When the laughter died McGraw's tone shifted to one of affection. "I watched him pitch, and I came to respect him. I watched him in the clubhouse and on the trains as we traveled and I came to enjoy him. I watched him marry and make a home for his wonderful Jane, and I came to a great fondness for him. I might insert here that he and Jane shared quarters with Mrs. McGraw and myself while their own apartment was made ready, and as the four of us are still on speaking terms there is proof of his humor and forbearance.

"Allow me to say, good people, that whatever I may accomplish in this game of baseball, whatever victories I may achieve and whatever glory may attach to my name, I shall always make it my proudest boast that I took this boy off first base and put him at the center of the diamond, which is his rightful place."

Yes, I thought, that is a proud boast, but what credit can obtain for recognizing the obvious? Less, certainly, than was Mathewson's due for suffering your arrogance and self-serving puffery. Impress this audience as you may; they have not heard you in your rage. How many would cheer if they'd suffered your profane persecution? You never threw a pitch, nor hit nor fielded one; only a small measure of this pennant is yours, and yet you stamp it with your own initials.

"Ten years ago, in Baltimore, on a similar occasion," McGraw was saying, "I voiced dismay at all the festivity, for I believed that the team could not survive in fit condition to play the Temple Cup series against . . . against . . . forgive me, but I can't remember who our opponents were that year." The joke went well among Giant fans. "This year, however, I have no such fear. My men could dance a reel for three nights running and still overwhelm the creatures of Mister Johnson's Western League." There was a stir of anticipation among the cheers that greeted this slur on the new league's ancestry. McGraw looked up at the first box, where the Giant president presided, and the two men exchanged nods.

"I say to you all, let the evening's joy be unconfined. The New York Giants have completed their season, and we have brought home to this great city the indisputable championship of baseball. I have no intention, Mister Brush has no intention, of entertaining the pitiful challenge of a bush-league operation. We will not entertain it, we will not acknowledge it"—there were gasps, then cries of encouragement and a rising ovation—"we will treat it with the contempt it deserves, and we will never dream of accepting it! Never!" McGraw shouted against the growing pandemonium. "Never! Never! Never!"

I sat on the Daimler's running board answering the stablehands' questions, insisting it was true: there'd be no series against Boston, McGraw had dismissed the challenge, there was war once again between the leagues. None voiced regret. What excitement might the series offer compared to the intoxication of the moment, McGraw and New York against the baseball world? Nor could it be considered a question of cowardice; none would apply that word to McGraw, and the Giants, runaway champions of the older league, were four-to-one favorites over the Bostons who'd eked out their title by the accidental measure of a wild pitch in their penultimate game against New York's new American League franchise, the Highlanders. No, it was McGraw's long-standing grudge against Ban Johnson and his works that lay behind all this, and these fellows of the stable were happily enlisted in Little Napoleon's battalion.

When Eli arrived he filled out the picture: the disbelieving commotion at the Erlanger, the parade up the aisle and out onto the Square with McGraw atop the shoulders of the throng, the battle among the reporters fighting for the telephones, and the hats that were passed in a spontaneous collection to compensate the Giant players for what they'd lose in failing to play (and win!) the World's Series. Given the emotion of the night the team might well come out hundreds of dollars the better when the kitty was divided—if it was divided.

Lost in the embroilment was our opportunity to produce World's Series rings, but for that Eli had a new plan. Rings celebrating the Giants' league title could be fashioned over the winter and presented on Opening Day of the 1905 season. Access to Brush, impossible in the tumult at the theater, could be obtained on the morrow. All that was needed was a design; how quickly could I render one? Tonight? Tomorrow morning? I asked about the cost; my voice was dull, and Eli wondered what was on my mind. I pleaded fatigue and told him to leave me at the corner of Nineteenth Street; I'd spend the night in the yet-unfurnished brownstone floors I'd leased for my home with Edith. Yes, I had a change of clothes there, and a blanket; yes, we'd talk in the morning. No, not now. When he pulled to the curb he reached to touch my shoulder, and I allowed it.

"Sport, you're acting like you've been to a wake. What's the trouble?"

"John McGraw."

"Mac? What's wrong with Mac?"

"He grates on me, Eli. I don't know that I want a ring of my design on his finger."

"Matty would have one too, you know."

"Yes, and so would Jack Warner. Eli, what's this about Warner showing our work?"

"How do you know about that?"

"He cut me out of the crowd at the theater. What have you offered him?"

"It's just an idea, sport. I had you in mind. I thought you might enjoy getting into more sporting designs. That ring of Matty's is so fine, everyone's thrilled with it. How would it be to do such rings on order for any player who wants one? Suppose a fellow goes four-for-four, or wins a big game with a hit or throws a shut-out. Warner could pitch him on the idea of buying a ring to commemorate the day. We could have an agent doing the same thing on every club some day. I'm getting to know enough of them to recruit a whole sales staff!"

"Eli, I won't have anything to do with Jack Warner in any regard, and that's that. Do you understand?"

After a pause, Eli patted my knee. "Cool down, sport, sure I understand. Look, Jack Warner isn't anything to me except a guy who plays a lousy hand at poker. He was feeling so blue one night—he'd lost his whole paycheck—and I just threw out the idea to cheer him up. I'm almost sorry I did it, it's the only thing he talks about when I see him. He's near the end of the trail, you know, and worries about what's going to happen to him."

"Let him walk to work with a lunchpail."

"What's that?"

"Nothing. I'd better go."

"Sport, if it means that much to you I'll drop Warner like a bad check. He's a good enough fellow, but you're my brother!"

"It's all right, Eli."

"But what do you think of the idea? An agent on every club?"

"Eli, I hadn't planned on doing a ring for Mathewson. The idea just came to me on the train that night. It was coincidence more than anything else. If I hadn't seen him in the parlor car I'd never have thought of it."

"But you did!"

"Yes, I did. Now why not leave it at that?"

Eli sighed. "I've really played this up with Uncle Sid. Putting a foot into the clubhouse of every city in the league would be quite something."

"Wouldn't it be too much, Eli? I'd end up having to design a special piece practically every day."

"Oh, you could just do a dozen or so drawings and give 'em a selection. Couldn't you?"

"And how would Mathewson feel if special commemoratives started popping up everywhere? No, I don't think so, Eli. This championship ring seems a good enough idea, but a slew of commemoratives . . . I'd rather not."

"Do me a favor, sport. Think about it a little while before you make up your mind."

"All right, Eli, I'll think about it."

"And do a proud job on these club rings. I'll need a drawing tomorrow."

"I'll see what comes."

"Attaboy. Good night."

I could not expect Eli to understand the nature of an offering. The ruby ring had been one; so, too, the several pieces I'd designed for Edith or at her request. These had won her father's respect, finally convincing him that I was a worthy suitor (though not without the added sanction of the firm's expanding success). It was another matter entirely to draw upon order. I was never proud of those designs nor flattered by what praise they won. Now I had an assignment that neither pleased nor excited me. McGraw had taken New York's night and made it exclusively his own; I feared that the city would cool to the club's achievement and recall only the manager's audacity, and I could think of nothing to celebrate this acid gesture. I sat on the floor of the empty apartment and stared at the bare wall, thinking it a sheet of sketching paper and constructing pieces in my mind's eye. I tried to look back beyond the images of the evening and recall the games I'd watched in April's cool shadows and July's encompassing heat. I saw Mathewson then, and with the memory of that perfection guiding my thought I worked to imagine something worthy.

I drew it the next morning at the shop. There was no sign of Eli until midday, when he charged in from a morning meeting with John Brush, grabbed the design, and hastened to Uncle Sid's corner office. Minutes later I was summoned to join them. My uncle was not convinced of the value of the good will we might accrue as suppliers of the Giants' championship rings: "Good will you can't eat." He was appalled at Eli's suggestion that the gemstones be diamonds rather than the sapphires I'd envisioned and lapsed into his native Russified Yiddish to express his opinion of diamond dealers. Why not sapphires? For that matter, why not amethysts? Our safe was full of amethysts.

"It should be diamonds," Eli insisted.

"By who, it has to be diamonds? This president?"

"No, Brush is leaving it to McGraw to approve the design. We're seeing him this afternoon."

"Who, we?"

"Jackie and me."

"Why me?" I asked.

"I need you to sell the design, sport."

"He'll want diamonds," I said. "Large diamonds."

"Why diamonds?" said my uncle. "Baguettes, chips you need, but I don't like dealing stones. Show him sapphires. Show him amethysts. If he insists diamonds—then we'll talk."

"You're tying my hands," said Eli.

"Good," said Uncle Sid.

Anticipating the need to excite McGraw's imagination, I called my cousin to the safe; we each spun our half of the combination and I withdrew the sample cases of sapphires and amethysts.

Eli had the Daimler warm at the curb, and we retraced our northward route of the previous evening, continuing up Broadway to Lincoln Square and then onto Columbus Avenue, where the new elevated tracks darkened the way. We parked at Eighty-Fifth Street. Here McGraw resided in a respectable apartment dwelling; here Mathewson and his bride lived as well, and Bresnahan. Ballplayers were gaining respect if landlords of such buildings would accept their signatures on leases.

The doorman announced us through a mouthpipe in the lobby and said "Mister McGraw's apartment" to the operator as we stepped into the lift. Eli hummed a tune; I squeezed the sample cases in my hand. The gate opened, the operator directed us to the right, and Eli rapped on the indicated door. A clubhouse functionary admitted us, took our coats, and led the way to what was less a parlor than a trophy room: bronze and silver plaques covered the walls, set among photographs and autographed pen-and-ink drawings of faces we'd seen at the Erlanger. Cups and statuettes filled two huge breakfronts that faced one another across a parquet floor. A fringed rug bordered the base of a black leather couch and two facing club chairs. Hanging plants blocked the window's northern view. It was a dark room, its floor cluttered with newspapers and telegrams. We were told that Mister McGraw would be with us shortly; we were not offered refreshments.

Suddenly he was in the room. His hair was parted slightly to the left of center, and slick with oil. He wore a vested blue suit over a high white collar and a diamond-patterned cravat. At his wrists he showed a large expanse of white cuff, held by the diamond-studded links he'd received the night before. The diamonds glittered no less than his eyes; stores of anger rested there. Success had not gentled him.

"Kappy! Good to see you," he said as he shook Eli's hand.

"Congratulations, Mac."

"What do they call you?" he said, turning to me. "Little Kappy?"

"No, no one's ever called me that."

"What, then?"

"Jackie, I suppose."

"You suppose! Aren't you sure?"

"Yes! Jackie."

"All right, Jackie, let's see what you've got in those cases."

Eli laughed. "Mac, you go directly for the loot. Don't you want to check the design first?"

"Design? Oh, sure, show me the design, son."

I took the sketch out of an envelope and McGraw grabbed it, walking to the window to examine it as the light allowed. "Is this the actual size?" he asked.

I looked at Eli, but he made no answer. "It's two-to-one," I said. "The ring would be half the size of the drawing."

"I know what two-to-one means!"

"Of course you do, Mac," said my brother.

"Now let's see these stones." McGraw snatched the cases from my hands and snapped them open. "I know these are sapphires, I have a sapphire stickpin. Cost me a hundred twenty dollars. What are these purple ones here?"

"Amethysts," I said.

"Amethysts! What the fuck is this, Kappy?"

"I don't know how high you want to go, Mac. As I told you when I called from Brush's office, we'll charge you for the stones and the gold and swallow the labor ourselves. If you're talking about a ring for every ballplayer——"

"And the coaches, and the batboys, and the clubhouse men, and the owners, and the ticket sellers. Everyone on this club is a champion. Every single person with this club is a world champion!"

"Well, you're talking about a lot of money then, Mac. Even if we stay with amethysts——"

"Fuck the cost. You're talking about world champions! What about diamonds?"

"Well, diamonds, Mac, that's a different league."

"It was diamonds last night! They all got diamond clips last night! I got these last night!" He held up his fists and shook them, displaying the links.

I was certain that my uncle would not countenance the purchase of fifty or more diamonds, even with a guaranteed return. I reached for the case of sapphires, took out the smallest, and set it on the lace that lay atop the table by the window. I placed a larger one beside the first and then another, larger yet. McGraw's eyes followed my hand, and I began to talk as he watched. "No matter what you do with a diamond in a ring," I said, "the diamond overshadows any meaning in the design. The question is always 'How big is the diamond?', never 'What is the meaning of the ring?' With a piece like this—" I pointed to the drawing he'd left on the table; McGraw glanced at it; I placed another sapphire in the row; his gaze followed—"there's a meaning beyond the gemstone or its cost, which is what everyone means when they ask the size of a diamond. This is not a diamond ring, it's a world's championship ring. Imagine it with the etching and engraving, think of the small chips here, here, and here—diamonds, yes, but small and integrated into the design without overwhelming it—and in its totality it says 'world champions of baseball.' It's true whether you use a diamond or a sapphire or a piece of polished quartz. It signifies a team, with two dozen ballplayers and, yes, the coaches and batboys and all the rest." I placed the last, largest sapphire in the row. "And the manager as well, I should add."

"Yes, indeed," said Eli.

McGraw studied the drawing, then looked at the sapphires in their array. "Damn, I think I know what you're getting at," he said, "but I can't quite see it."

"Do you know the piece we did for Mathewson to mark his no-hit game?" I asked.

"That's yours? I've seen it, he wears it all the time, but I never took a close look."

"If you study it I think you'll understand what we have in mind."

"Right. Let's go." He strode out of the parlor, leaving us startled. "Come on," he shouted back at us, "we're going to have a look at Matty's ring."

Eli followed instantly; I didn't move. Whatever my resentment of McGraw, the chance to do the rings was sufficient reason to beard him in his home. An encounter with Mathewson was quite another matter. I had no wish to draw closer than the distance which had separated us at the Erlanger.

Eli reappeared at the door. "Come on, sport!"

"No, I'll stay here."

"The devil you will."

"Eli, there's no need——"

"There's every need! God of light, Jackie, what am I supposed to tell McGraw! I can't explain this thing you've got over Matty, he'll laugh us out of here! Now, come on!"

I collected the jewel cases and the drawing and followed Eli. In the hallway McGraw had his finger pressed to the elevator call button and kept it there until the lift arrived. He snapped "Five!" to the operator and rapped his knuckles against the wall as we descended. When the gate opened he walked quickly down the hall, taking a ring of keys from his vest pocket. He inserted one in the lock of a polished wooden door, pushed the door open, and marched ahead.

"Matty! Where are you? Jane! Matty!"

We were in Mathewson's salon. There were some packing crates on the floor, the walls were bare, and the furniture was covered with white cloth. There were no drapes about the windows, though there were brass poles in the supports, and sunlight streamed in from the west.

"Matty! Jane!"

"She's out, Mac," came a voice, and Mathewson emerged from an interior hall. He was naked, save for a white towel draped about his shoulders; his hair was wet, and droplets of water gleamed on his pearl-pink skin. Momentarily surprised at seeing the two of us with McGraw, he drew his muscles taut, and his whole body rippled with power, water sparkling in the rays of the sun. Then he relaxed, his weight falling on his right leg while the left bent slightly, and he smiled.

"Hello, Kappy. Hello, Master Kapp. Excuse me while I put something on."

"Bring your ring, the no-hit ring!" McGraw called after him. Seeing the expression on my face, McGraw's eyes flashed. "You never saw such a body, did you?"

Not of flesh, I thought; once, in marble.

"Give me a body like that," said McGraw, "and I'd have been twice the player I was."

"Jackie was a pitcher himself, Mac, and a good one. He had an offer from the New York–Penn League, but he came up with a bad arm. Right, sport? Sport? What's the matter?"

"Nothing. I wasn't expecting . . . never mind."

"Lefthander?" McGraw said, sizing me up with a professional eye, and I nodded. "You're small," he said. I was three inches taller than the manager. "Did you use the breaking stuff?"

"That's what did in my arm."

McGraw turned away, his interest ended. I watched the hallway for Mathewson while Eli took a seat on a cloth-covered chair. "You know, Mac," he said, "I have to admit I'm damned sorry we won't get to see you go against Boston. As a baseball fan, that is."

"And maybe you'd have had something riding on it, just to make it interesting. Hey, Kappy?"

"Maybe a little something, Mac."

"You're full of shit, Kappy."

"I guess I am, Mac."

"Fuck it. Fuck them all. Fuck Boston and fuck the commission and fuck Ban Johnson. I told that bastard I'd get even. You know he was going to pull the Orioles out from under me and move them here? Leave me with shit on my chin in Baltimore? Of course you know it, anyone could see it coming, anyone. He was dying to get into New York, and to hell with John McGraw. He tried to destroy me. He put the umpires after me, he put everyone after me, he made it open season on me. Look at this. Look at it!" McGraw pulled up his trouser leg to reveal a great white scar running from his ankle to his knee. "Who do you think did that to me? You think Dick fucking Harley put that there? Hell, no, it was Johnson! Harley got five hundred dollars for spiking me! And what did I get? A suspension! I was on crutches, and couldn't even issue myself a paycheck."

He let the trouser leg fall and paced a rapid circle in the center of the room, his clenched fists punching at the air. "They all came after me. The word was out: 'Get McGraw!' Then move the club to New York and leave me licking my wounds in Crabtown. Well, fuck him. Let him have his miserable franchise in New York. I got here first and I'll never let them beat me, never! I'm just damn sorry his New York club didn't win their flag. I'd have rammed it right down their throats. Ban Johnson, Byron Bancroft fucking Johnson. He can take his Western League and his National Commission and his World's Series and shove them all up his big brown asshole, if he can tell it from his face, which would be some trick. And that goes for all of them! We're champions of the world! We're——"

He stopped, looked past me, and moderated his tone. "We're champions of baseball, no question about it. Right, Matty?"

"Champions, Mac." Mathewson stood at the head of the hallway, attaching a white collar to his gray and white striped shirt. He'd put on gray slacks and lustrous black half-boots; his hair was brushed and parted far on the left. Finishing with his collar, he walked to me and offered his hand. "It's a pleasure to meet you at last," he said in a rich baritone. "I hope you weren't discomforted."

"I hope you weren't." His grip was strong. He let go of my hand, acknowledged Eli with a nod, and went to McGraw, extending the ring at the manager's eye level. "Here it is, Mac."

"Yes, I see," said McGraw. "Nice work. What did they charge you for it?"

"It was a gift."

"A gift! Jesus, Matty, don't you ever pay for anything?"

"It's Master Kapp's own design," Mathewson said. "His cousin worked the ring, and his brother brought it to me on the day I won my twentieth game, three years ago. I understand it meant a great deal to Master Kapp, and it means a great deal to me."

"Thank you," I said.

"Is that another design there? May I see it?"

"Of course." I gave Mathewson the sketch. "The stone would be a sapphire."

"Yes, that's excellent. I look forward to having it."

"There'd be one for everyone on the club."

"Of course."

"I thought diamonds would be better, Matty," said McGraw.

"If Master Kapp designed it for a sapphire, a sapphire would be best."

"Tell you what, Mac," said Eli. "I'll make one with a diamond for you, and we'll do the rest in sapphires. How's that?"

"Yes! You do that. Matty, do you want a diamond?"

"I'd prefer it as it's drawn. I like it very much."

"Thank you," I said again.

"I wonder if I might have this drawing when you're done with it."

"Of course. Take it now, I can do another."

"You're sure?"

"Yes."

"I think your work is extraordinary."

"So is yours."

McGraw made an impolite noise. "You two sound like Mistress Mary's School for Girls," he said. "It's a good thing you don't pitch like that."

"Far from it," said Eli.

"Yes," I said. "I remember the way you put Padden on his back in St. Louis."

"A two-strike pitch," Mathewson said. "He was leaning in for the fadeaway. The bat landed on his head."

"Yes, it did!" We laughed together.

"Well, Mac, do we have a deal?" said Eli, rising from his chair. "How many should we make?"

"I'll give you a list. I want every one initialed."

"We'll see to that."

"And we'll hand them out on Opening Day. I'll see to the ceremonies." There were handshakes all around.

"Would you do something for me?" Mathewson asked at the door. "Would you sign this?" He gave me the sketch, and I borrowed his pen to write the date and my signature. "I appreciate this very much," said Mathewson.

I could think of nothing to say.

Edith and I were married on the sixteenth of October and sailed for Europe the following day. Upon our return we found the Nineteenth Street apartment strewn with gift packages, and we passed New Year's Day opening them and writing grateful acknowledgments. I was astonished to find in the collection a present from Mathewson, a quilted comforter. His note wished us the happiness he'd been vouchsafed in his own marriage, and there was a postscript in his wife's hand which told us that the quilt had been fashioned in a nearby Quaker community. Put to its purpose that winter, it proved the most useful and intimate of gifts.

In late January we had my in-laws to dinner. When Mrs. Sonnheim complained of a slight chill Edith fetched the comforter and spread it on her mother's lap. Mrs. Sonnheim remarked on the decorative stitching, an intricate interweaving of signs, and she guessed that these were a series of hexes. Mister Sonnheim allowed that this was an apt gift from a ballplayer, as they were notorious for superstition, but he wondered if Mathewson wasn't outside that run. He'd heard something of the man from colleagues in business, he said, and knew of a young initiate in his own firm who'd been a classmate of the pitcher at Bucknell. That Mathewson's name and character were known to my father-in-law struck me as a signal of his accumulating celebrity; Mister Sonnheim had always strongly differentiated between the game he'd enjoyed as a young man and the play of the professional leagues, which he held beneath notice. Now he quizzed me about Mathewson, and I took the opportunity to ask him to be my guest at the Polo Grounds on Opening Day. I added that our firm had a role in the ceremonies, coloring my attendance as a matter of business. He answered that he would have to consult his appointment book. I received his written acceptance the next day, addressed to my office.

He could hardly have escaped the mid-April celebration. It was a citywide festival: Fifth Avenue and Broadway were festooned with pennants and bunting, and an automotive procession, the city's first, began at noon at the Grand Army Plaza south of Central Park. Patrons of the team had donated their cars—Eli had been quick to offer the Daimler—and nearly fifty such ornaments were in line to take the players south to Washington Square, east to Cooper Union, and then all the eight miles north to the Polo Grounds. The way rang with cheers for McGraw, alone in the premier car, and for the athletes in uniform who followed in sets of twos and threes. The rear was taken up by the visiting Boston Nationals, who accepted the plaudits of the citizens in grand pantomimes of the Giant heroes. Forty thousand awaited the teams at the ballpark; the ever-growing grandstand could hardly accommodate that number, and the standees were roped into corrals that bordered the whole playing field, with a narrow gauntlet cut from each dugout to home plate. A regimental band tootled and drummed, and flags bristled over the grandstand roof in the stiff breeze. When the line of autos drove onto the field the mob broke through the unstable barriers to lay hands upon the heroes and raise them to their shoulders.

In our box seats in the front row of the grandstand's upper deck my father-in-law observed all this with amused vivacity. He'd confessed that to his mind the words "Polo Grounds" still connoted publisher James Gordon Bennett's private field on upper Fifth Avenue, and remembered that the Metropolitan professionals had had to wait their turn while amateur sportsmen represented private clubs in gentlemanly competition. He delighted me by recalling that he'd once shaken the hand of Alexander Cartwright, the grand doyen of the New York Athletic Club who'd encoded the rules of the New York game before the Civil War. When the clubs took their warm-ups Mister Sonnheim noted with interest that all the players wore mitts, and wondered how long that had been the practice. I said that prizefighters had also taken to wearing gloves, and he raised an eyebrow. He was astonished at catcher Bresnahan's get-up of mask, chest protector, and shin guards, and accepted my explanation that the gear was necessary now that catchers crouched directly behind the plate; in his day they'd received the pitches a safe twenty feet back.

The occasion was freighted with ironies and absurdities. Monte Ward was there, who'd managed the Giants to the Temple Cup victory over McGraw's Orioles a decade before. Harry Pulliam was there, the National League president; he'd swallowed a good deal of pride to attend, for his position had been very nearly wrecked by McGraw's refusal to play the World's Series. Jim Jeffries, the heavyweight champion, was greatly hailed, and when he took McGraw's arm and raised it high in the classic victor's pose the difference in their sizes was comic. And the politicians, of course: Mayor McClellan, Hayes of the Fire Department, McAdoo of Police, Woodbury the Street Commissioner. Scampering among them was Eli, in top hat and morning clothes, missing not an ear for a confidential whisper nor a back to slap.

The players formed ranks along the basepaths, and I borrowed Mister Sonnheim's opera glasses to see their faces. Two stirred my memory: Warner, who'd managed to hang on for another year, and Strang, returned from Boston via Brooklyn and slated for utility duty. There'd be no ring for Dixie, at least. The prizes, in small boxes of blue velvet, were arrayed on a table at home plate, and when the presentations began the line resembled a bucket brigade: Eli would hand one to Woodbury, who'd hand it to McAdoo, then McAdoo to Hayes, Hayes to Pulliam, Pulliam then to the Mayor, and finally to the ballplayer who'd trotted out at the call of his name. I saw that every Giant save Mathewson snapped open the box and examined the ring immediately; the pitcher slipped his into a hip pocket without a glance. But then, he had the sketch. With the players called alphabetically Warner was the last, and when he returned to line he teased the unrewarded Strang.

After the players, McGraw. The Mayor paused to note that of all the rings only McGraw's held a diamond, and he called this "a fitting tribute to a great leader of men, an inspired decision on the part of the makers of the ring." I laughed, and saw that Mathewson laughed as well.

He was not the assigned pitcher; McGinnity had the call. He'd thrown a hundred more innings than Mathewson in 1904, a thousand more pitches. McGraw called on him with a day's rest, or twice in a day, as if eager to strain the last use out of a mighty machine before its inevitable breakdown. Always he threw with the force of a driving piston, and although hitters choked up high on their bats and chipped at the ball with half-swings it was like trying to tag sparks as they shot out of a fire. Boston could do nothing at bat, but New York created electricity with its bunts, sacrifices, steals, and hit-and-run plays. All this was new to Mister Sonnheim, and he was fascinated with the dynamism of the style. McGraw varied his tactics: the Giants scored twice in the first inning with bunts and daring baserunning, slugged out three more runs in the third, and reverted to inside play in the fourth to push their lead to eight runs. It was ten to nothing after six, and McGinnity was grinding up the visitors, yet my father-in-law made no suggestion that we leave. He actually groaned in disappointment when Boston ended the Iron Man's shut-out in the eighth, and asked if it were possible that Mathewson was a superior pitcher.

All the scoring made it a lengthy contest; dusk approached and the grandstand was but three-quarters full when McGinnity came out to pitch the ninth. Eli, who'd watched from the club box beside the Giant dugout, arrived to pay his respects, looking positively ambassadorial in his get-up. Mister Sonnheim greeted him formally and kindly refused his offer of a ride downtown in the Daimler, pointing instead to the outfield perimeter where his coach and driver waited in the line of livery. For me, Eli had an invitation to a celebration dinner with the team at Delmonico's. I begged off as a married man.

"I bet you wouldn't say so if Dad Sonnheim weren't here!" Eli joked. Dad Sonnheim! I shuddered at the phrase.

"You bet frequently, Mister Kapinski?" said my father-in-law, more as a statement than a question.

"No more than you do at the Exchange," said Eli.

"My enterprise is the calculation of risk," said Mister Sonnheim. "The capital belongs to the clients." He pulled on his gray gloves. "Jacob, I thank you very much for the afternoon. I look forward to an opportunity to provide you with an equally enthralling entertainment."

"It was entirely my pleasure, sir."

"Good afternoon. Good afternoon, Mister Kapinski." He departed after handshakes, and Eli slipped into his seat.

"Okay, now that the coast is clear, are you coming to Delmonico's?" he asked.

"No, I'm going home."

"Say, sport, that gal must really have a sign on you."

"She does that," I said cheerfully.

"Newlyweds! There's no talking to them. Any messages for Matty, then?"

"You can tell him how much we appreciate his gift."

"I'll bet you do! Over it, under it, all around the——"

"Eli!"

"Sorry, sport, but one night out might add a little to your life. Come with me, there are some good fellows on the club."

"Eli, spending an evening in the company of Jack Warner and Dixie Strang is not my idea of adding something to my life."

"Oh, they've gotten over that years ago."

"Over what? Jew baiting?"

"You've got to look at it from their side, sport. They didn't know what you were about in Chicago. Neither did you, for that matter."

"I know now, and I'm going home tonight."

"Newlyweds." Eli shook his head. "Well, I'll miss you, nevertheless. Terrific game, wasn't it? I had fifty on it. I'll have fifty on Matty tomorrow, too. Fifty any time it's Matty or the Iron Man."

"And another Daimler in September?"

Eli grinned. "It's a sure thing, sport. Only a fool would pass on a sure thing."

Mathewson won the next day, and kept on winning; McGraw's courtship of the mob continued apace. In New York it was a love affair, a passion that increased in proportion to the distance between the parties. In the box seats the manager was admired, in the grandstand lionized, and on the bluff worshipped. In other cities he superseded the King of Spain or Filipino *ladrones* as The Enemy. In Boston it required barricades for the Giants to safely reach the ballpark; in Pittsburgh the police refused their protection and the club had to run an angry gauntlet to the clubhouse; in St. Louis a hotel denied them accommodations, claiming that they couldn't guarantee the safety of other guests. About the Giants they didn't seem to care. McGraw booked the team into the Chase, and from that time forward they put up in the hotels that Eli favored, though they had to pay their bills in advance. In June the National Commission suspended McGraw as a threat to the public safety; he sued the commission, and won. In spite of it all—because of it, some argued—the team won twice as often as it lost, and drew twice the crowds.

There'd been a moment in Mathewson's apartment when I'd

seen the player's effect on the manager; he'd dropped his voice and abandoned profanity. Now I saw signs of an opposite influence, as when Pittsburgh came to the Polo Grounds for a key series. That powerful club, which the Giants had deposed as league champions, was anchored at shortstop by the finest ballplayer in the land. Honus Wagner matched Mathewson for size, and in the infield he stood like a gnarled oak with bowed roots, his long arms branching nearly to the ground; with his oversized hands he'd scoop up anything hit to his enormous range, gathering with the ball a large measure of infield dirt, and he would fling the whole package toward first base, debris trailing off like a comet's tail, the toss ever straight and true. With an adequate supporting cast Wagner could steal a whole season, and his team had been strengthened with the addition of a collegian named Mike Lynch—such was Mathewson's example that more and more pitchers were elevated from the college ranks. In the youngster's debut against the Giants McGraw tested him with a shower of glorious invective, and Fred Clarke, the Pittsburgh manager, responded protectively, striding in from left field to strike a fistic stance before McGraw that would have done credit to the Nonpareil. The umpire defused the crisis by ordering McGraw to the clubhouse. Little Mac took some minutes to oblige while the crowd howled its outrage. Finally a grumbling peace was achieved and the game proceeded, only to be upset the very next inning when Mathewson, apparently deputized by the banished McGraw, charged out of the dugout to instigate a new imbroglio. Whether he was in earnest, as appeared the case, or acting out a delicious mockery of both managers was a matter of conjecture, but the umpire bid him join McGraw in the showers. The Giants lost nothing by the ejection, for he was not the pitcher of the day, and Mathewson gained a few hours vacation, but it was not a happy occasion to witness, and I cursed McGraw for a hellion.

Some weeks later there came a report from Philadelphia that I could scarcely credit. A fight had broken out between players—not unusual in itself; McGraw had led his entire squad onto the field to join the fracas—standard behavior by then; spectactors had spilled out of the grandstand to enter the battle—not unprecedented; and when the fray was finally extinguished there was a "lemonade boy" (so described by the local press) with a split, bleeding lip and some loosened teeth, and dozens to swear that it had been Mathewson who'd struck the blow. The papers carried no denial by the pitcher and a defense of the act by McGraw, who seemed delighted that the veneer of civilization had slipped from the athlete. I didn't know what to make of the story, and wished that there had been New York writers present to balance the reportage.

I asked about the incident when we arrived in Philadelphia to begin our western swing, but a local client made light of it, reminding me that a sportswriter's dedication to objective truth was something less than fervid and that the least believable fancy often made the best story—especially if it might lead to the suspension of an unhittable foe. That evening the client called back at our Bellevue suite with a late edition of the *Bulletin* and slapped it into my hand with a wide grin.

"Take a look at what your man's done now!" he exclaimed, and I scanned the sports pages expecting the worst. Instead I found this headline:

"MATTY" PERFECT
IN CHICAGO

Allows Cubs Neither
Hit nor Walk in 1–0
Victory—Two Giant
Errors Mar the Game

I called the desk to send a wire:

CHRISTY MATHEWSON
NEW YORK GIANTS BASEBALL CLUB
WEST SIDE GROUNDS
CHICAGO, ILLINOIS

SINCERE CONGRATULATIONS ON A MAGNIFICENT PERFORMANCE
YOUR REWARD SHALL BE NO LESS THAN THE LAST
J. KAPP

The following afternoon, as we checked out, the desk clerk handed me his answering telegram.

J. KAPP
BELLEVUE HOTEL
PHILADELPHIA, PENNSYLVANIA

A DOUBLE BLESSING IS A DOUBLE GRACE OCCASION SMILES UPON A SECOND LEAVE
MATHEWSON

On the train to Pittsburgh I wrote two letters. One went to the shop, with instructions to fashion a ring identical in form to the first but to substitute an emerald for the ruby and to have the engraving read *Chicago/ June 13, 1905*. I had the cost charged to my own account, consulting Eli not at all. The second letter was to Edith; I enclosed the telegram and asked if she could identify the phrasing, which I'd guessed had a classical origin. Her letter was waiting for me in St. Louis. The lines were from Shakespeare; a vigorous youth spoke them to his father, a foolish old figure, in *Hamlet, Prince of Denmark*. Edith advised me not to take the characterization to heart, adding this citation from another play: "The devil can cite scripture for his purpose." These literary games only added to my confusion. I thought on the matter all through the swing, and finally chose against taking his wire as an invitation to a second meeting. Fearing that a close association would tarnish the figure that seemed so prodigious from a distance, I found it enough to be, in Eli's phrase of long ago, a worshipper from afar. When the new ring was done I had it delivered by messenger to the Polo Grounds.

Mathewson's note arrived the following day:

August 9, 1905

My dear Master Kapp,

Once again I am honored to receive a model of your artistry. With luck I may yet claim the city's largest private collection of your work.

You may know that Richard Sonnheim has invited me to dine with you at his club on the ninth of September, after his return from his summer residence. I look forward to the occasion with enthusiasm.

I am, in deep appreciation,
Yours,
CHRISTY MATHEWSON

I had my father-in-law's invitation for that date, but it made no mention of Mathewson. This, then, was his return favor for our Polo Grounds excursion. I couldn't fathom Mathewson's eagerness for my company. Perhaps he was merely polite, perhaps unduly impressed with my craft; perhaps he had from Eli some impression of my feeling for him and wanted to examine with his own eyes this strange specimen of idolator. Yet I could see nothing but to accept Mister Sonnheim's offer and to feign surprise when he revealed the arrangements.

I'd been to his club once before, for breakfast on the morning after I'd asked for Edith's hand, to hear his answer. I thus associated the place with a high degree of anxiety, and the appointed evening in September had nothing to reduce the fever. I was late; unable to find studs for my shirt at home, I had to race back to the shop to lift a pair from a sample drawer. I arrived at Stuyvesant Square perspiring in the warm twilight, decided that another five minutes' tardiness could do no further harm, and took a slow walk around the block to settle myself. I bought a late edition of the *Sun* at a kiosk and checked the scores. Mathewson had beaten Boston for his twenty-eighth victory that afternoon, and the Giants led the league by twelve games.

In the huge and opulent members' lounge Mathewson towered over a circle of admirers. He was no less at ease in black tie than in his playing togs or the formal dress he'd sported at the Erlanger. He did not so much fit into his surroundings as define them; he seemed innately, magnetically right in every circumstance. My father-in-law was at his side, and catching sight of me he left the pitcher to greet me, shrugging off my apologies and declaring his delight at the scene Mathewson had engendered. Grover Cleveland had never created such a fuss! There was a table reserved in the third floor dining room; would I care to proceed there? He'd disengage Mathewson and follow shortly.

Our table setting was ornate, and I passed time studying the patterns on the crystal and china and eavesdropping on diners close at hand. A footman offered a glass of sparkling wine, which I downed swiftly and regretted immediately as a damp sweat broke out on my brow. At last they entered. My father-in-law, a tall and impeccably groomed gentleman of dignified bearing, was dwarfed by the athlete. Heads turned, a man at a nearby table began to tap his knife against his glass, and others took up the tribute. Some stood to applaud, and the ovation became general; it did not cease until Mathewson had waved to every corner of the room and taken his seat beside me. I saw the new emerald ring on the fourth finger of his right hand. He inquired after my wife, and I returned the courtesy. Mister Sonnheim remarked on the extraordinary reception he'd received, and the pitcher made a deprecatory comment; it was the team's success, not his own, that deserved the applause.

The conversation touched on such polite matters as mutual acquaintances in Philadelphia society and the comparative delights of that city versus New York. I listened in silence, nodding when, as happened from time to time, Mathewson turned to me and asked, "Don't you agree?" or "Do you think so?" When a footman asked if I was done, I looked down and saw an untouched plate of thinly sliced cold meats before me. I nodded, and he removed the plate and served the chowder. Mister Sonnheim began to recount his memory of Alexander Cartwright, whose rules had given baseball a foundation that yet survived after sixty years of innovation and experiment. Mathewson was fascinated.

"Would you say that Cartwright's mind had a particularly mathematical bent?" he asked.

"Decidedly so," said our host. "He once defined his endeavor as an attempt to balance the arithmetic of the game against its geometry. That was his phrase: to balance the arithmetic against the geometry."

"What a success he had! Don't you agree?" I nodded; Mathewson continued. "All those balances—so exact, so demanding and tantalizing. Nothing in the game is easy, yet nothing is impossible. It's a game of intricate simplicity."

"And as such, resoundingly American," said Mister Sonnheim.

"We've whole systems of intricate simplicities. Our economy, our politics, our entertainments, all so reflective of the national character. I wonder if you're familiar with the popular passion of the Afghani tribes, as a contrast?"

"Bushkazi," said Mathewson.

"Exactly. Jacob, do you know it?"

I confessed my ignorance.

"It's played between opposing squadrons of horsemen, as many as fifty on a side. The contest is for possession of some barbaric thing, I've forgotten what."

"A decapitated goat," said Mathewson.

"Just so, a decapitated goat. Unfortunate; we're to have leg of lamb tonight. Had I remembered I'd not have raised the subject. But never mind. The object of the game is to advance the goat across a field almost two miles long, to circle a post at that extension, and to return to the starting point. In order to wrest the ball—the goat—from the horseman in possession, the opposing riders may commit almost any form of mayhem upon him or his fellows. I'm told it's considered a rather poor match in which several of the horsemen aren't killed outright."

"It certainly puts baseball's use of the term 'sacrifice' in a different light," said Mathewson, grinning.

"Rather," said Mister Sonnheim. "The game—if game it be—certainly derives from the fight for the spoils of the hunt. The motions of the horsemen, obscure and random to the unpracticed eye, were highly praised by the cavalry officer from whom I had this information. Clearly it's battle in a most ancient form. The game permits the expression of the innate aggression that lies within the tribesman's breast. Games of possession, games of targetry—they're all as obvious as *bushkazi* in their practice of hunting or warmaking skills. How civilized is baseball in comparison, how subtle, how refined! Baseball, it seems to me, makes little sense in those other terms. It is the most intellectual of the physical sports. It is totally artificial, creating its own time, existing within its own space. There is nothing real about it."

"Except the men who play it," said Mathewson. "And more than that, it occurs to me. After all, what are the two physical acts most basic to the play? Throwing and clubbing. What could be more ancient? Certainly they predate the Afghani's mastery of the horse. If we accept the compelling arguments of the naturalists, we have to grant that our prehistoric forebears employed those same arts against the creatures of nature—indeed, against one another. Even in holy writ, mustn't we imagine that Cain slew Abel with a stone guided by the bare hand, or a club wielded as a bludgeon? Think of it. I stand on the pitcher's mound, the batter at home plate. We are surrounded by every manifestation of civilization: the manicured field, the rising grandstand, the railway beyond the outfield, the buildings on the bluff. Yet my action in throwing and his in swinging are echoes of the most primitive brutality.

"Did Cartwright say that his endeavor was to balance the arithmetic of the game against its geometry? All of sport, from *bushkazi* to baseball, is man's endeavor to balance his animal instinct against his civilizing intellect. On the sporting field, to borrow Mister Disraeli's phrase, we are both ape and angel. Don't you think so?"

The chowder was gone, a salad in its place. "I remember a speech at an awards dinner years ago," I said. "Jack Warner gave it, as a matter of fact. The whole point was that baseball and life made the same demands. 'Succeed in one and you'll succeed in the other,' he said. He made the point repeatedly."

Mathewson laughed, and I was thrilled. "I know the speech," he said. "Jack offered it to me for five dollars when I joined the club. All the fellows use it, or something close. 'Baseball is a lot like life,' followed by ten pages of analogy. Is that the one? Of course." He laughed again. "Well, I paid him his five dollars and sat down to study the speech, and do you know? I found practically every sentence contrary to my own belief. After all, baseball isn't anything like life. I think that was your point, sir, when you said there's nothing real about it. In that sense I agree. In truth, nothing in the game appealed to me as much as its unreality. Baseball is all clean lines and clear decisions. Wouldn't life be far easier if it consisted of a series of definitive calls: safe or out, fair or foul, strike or ball. Oh, for a life like that, where every day produces a clear winner and an equally clear loser, and back to it the next day with the slate wiped clean and the teams starting out equal. Yes, a line score is a very stark statement, isn't it? The numbers tell the essential story. All the rest is mere detail."

"Ah! Here's the lamb," said Mister Sonnheim as the footman served.

"I hope you've regained your appetite for it, sir," the pitcher said.

"I'm assured it's a domestic cut, not the afterleavings of a *bushkazi* match," said our host. "You'll have some wine?"

"Are the members discreet?"

"Absolutely."

"Then half a glass, please." He smiled. "I have something of a reputation to preserve."

"Would your reputation suffer if you were to join our table for bridge after the meal?"

"I'd be delighted."

"And you, of course, Jacob."

"I think not, sir. Thank you all the same."

"You leave it to your brother to uphold the family's honor at the card table," said Mathewson.

"I never thought of it in terms of honor," I said.

"Cards are no less honorable a game than any other, and like wine they reveal a man's character. It all comes out under the pressure of the play. I always knew Kappy to retain his honor at the table, if not his stake."

"You play with him regularly?" asked Mister Sonnheim.

"Not for some time. I gave up my chair at the Victoria when I married. Interesting about your brother, he never bluffs. He plays the hand as it's dealt, and never pretends otherwise. Is he like that in other circumstances?"

"I never thought to notice."

"He enjoys the play, but he has no capacity for pretense. I find him delightful."

"Bit of a black sheep, I should think," said Mister Sonnheim. "I gather that his gambling hardly ends at the card table."

"But cards are not gambling, surely," said Mathewson.

"Sir?"

"Certainly not. A competition for stakes is a far cry from gambling. God forbid that I should gamble; I find it atrocious, and its practitioners are the worst of men."

"But poker—" I began.

"—is a competition for stakes, a calculation of risk. The man who thinks of poker, or any card game, as a matter of blind fate will surely end a beggar. The play of cards is a matter of continual calculation, and he who best judges the odds will win the stakes. So also chess, or checkers. So also a professional baseball game, for that matter. We play for championship stakes, after all."

"You're quite sure that this is not a semantic quibble," said Mister Sonnheim.

"Sir. You and I both breed horses. I consider that my horse can run a mile faster than yours. We lay stakes upon the argument, we select jockeys to ride them—another matter of our own judgment—and we set them on the course in head-to-head competition. The man who's bred the best horse wins the stakes. That is not gambling. Here is gambling: a third and fourth party at the rail, who have nothing to do with the breeding or the training of the animals, lay a bet on the outcome of the race. And perhaps —more than perhaps, inevitably—one of those bettors seeks to ensure his investment by taking the jockey aside and suggesting that it would be worth his while to slip a stirrup at the three-quarter mile. It is, I say, the certain consequence of gambling to reduce the competition to the gain or loss of money. Exercise, honor, fair play, achievement—all go by the board. If the sole importance of the enterprise is the money bet upon it, then it must inevitably end in a situation where the competitor has more to gain by losing than winning. And that is atrocious."

"It is indeed," said Mister Sonnheim. "I cannot argue the point. And yet it does seem that the third and fourth parties are each pitting their best judgment against the other—calculating the risk and competing for the stakes, as it were."

"The point remains that they are third and fourth parties, with no direct involvement in the competition," Mathewson answered. "I cannot see any dignity in their occupation, and I fear their influence."

"Well might you do it," said our host. "We'll allow no third and fourth parties in the card room tonight."

"I thank you, sir." He raised his glass. "To our host."

"To your continued success, Mister Mathewson," said my father-in-law.

"And to a world's championship," I added.

"A world's championship," Mathewson repeated, and sipped the wine. "I do wish that Mac had taken up the challenge to play Boston last year. I hope he'll rectify the matter this year. I've told him how much it would mean to me."

"I've already done a sketch," I said.

"That's a ring I look forward to wearing. You'll do it wonderfully, I'm sure. Do you know this one, sir? It's a duplicate of another, four years old, with a different gemstone."

"It's very handsome, Jacob."

"The emerald works quite as well as the ruby," said Mathewson. "Together they're a handsome pair. You've no idea what that first ring meant to me. It restored the moment, gave it an enduring character. What a moment it was! The ground ball to Dixie, the throw to Ganzel—I felt it had been worth the trial after all. It wasn't easy to sign a major league contract, you know. My parents were against it. It wasn't a fit career for a gentleman, after all. They regarded my earlier play as simply another form of summer employment, but the major leagues! It just wasn't done. When I signed with the Philadelphia club they were aghast. Our family was known in Philadelphia society, and to have me playing with the ball club! I hadn't realized it would wound them so deeply. I explained this to Mister Mack, and he was quite gracious about it, recommending that I go to New York. That, and a promise that I'd never play on Sunday, had to suffice for my parents. I felt I must test myself at the highest level of play. The no-hit game was the peak, the last out intensely fulfilling, and yet within an hour it was fading, within a month it was a distant memory, and by season's end, with the club in the cellar, it was almost forgotten."

Still looking at my father-in-law, Mathewson gripped my wrist with his left hand. "After my last turn of the season, when I'd won and it hardly mattered that I'd won, when I was giving some considerable thought to whether I'd play again the following year, out of nowhere came this marvelous piece of work. It made the moment real again, and I never wear it but that I feel it anew. I don't know that even a world's championship ring would mean as much, but I certainly pray for the chance at one."

"And that done?" said Mister Sonnheim. "Would you then retire?"

Mathewson smiled. "Alexander wept, having no more worlds to conquer. In the game, a new world begins every spring." He touched the ring. "And this, after all, is an individual achievement, whereas a world's championship is a team title. I echo McGraw, who saw this in the clubhouse today and told me that I might end with a dozen of them, but they wouldn't mean a thing—not his expression, of course—compared with a title."

"He'll play the Series this year?" I asked.

"There's no telling. Certainly I hope so." He took another sip of wine as the footman collected the plates and set out a bowl of gelatin. "You have an aversion to the man, don't you? I sensed it at my apartment."

"I can't imagine how you put up with him."

"On the contrary, he's the most compelling personality I've ever encountered. Alexander Pope had him exactly: 'A being darkly wise and rudely great.' You must realize that professional baseball is thick with men who share his vices but possess none of his virtues. And his virtues are real. His loyalty to his players is absolute. He never scolds for a physical error, however much he rages at a mental one. He encourages initiative, and makes certain that every man knows what's expected of him. And his knowledge of the game—why, you know as well as I that he practically invented modern baseball with the Orioles."

"Not alone."

"No, but certainly foremost. Do you think we'd win without him? McGraw is the team, the players are his marionettes. His hands are calloused from pulling the strings."

"I can't argue his expertise," said Mister Sonnheim, "and certainly I found his style of play most energetic and effective. But he seems an abrasive personality, even an excessive one."

"What successful personage is not excessive in one regard or another?" said Mathewson. "Understand the man, if you can. He is not like you or me. He's certainly unsubtle. One knows his mind immediately."

"It wasn't so much his mind that put me off as the venom of his expression," I said.

"About Ban Johnson? Mild, by his usual standard. But consider two things about him. First, the day of your encounter. He'd merely upset the entire establishment of baseball. At a stroke he'd overturned two years of delicate negotiation and intricate financial arrangment. He'd been up all the night before with the Commissioners, fending off their threats and turning down their bribes. No doubt it gave him enormous satisfaction, but it's not the sort of thing that can be done twice. That's why I think there'll be a World's Series this year. And consider too his roots. He didn't have our advantages. You grew up in this great city, with access to all its treasures. Twice a year I had extended visits to Philadelphia. What has Olean, New York, to offer the energetic mind? I chose baseball from a range of other options. Had I never pitched an inning, there were still great worlds open to me by reason of my family and upbringing. And you—though your arm failed you, you've found another realm for your expression, you travel, you have some degree of economic security. For McGraw it was baseball or nothing—nothing, that is, but the mean soil of a Cattaraugus County farm. For McGraw baseball wasn't one way out, but the only way out. He had nothing but baseball, and he still believes that, absent baseball, he is nothing. Therefore, what threatens him in baseball threatens him entirely, and what you or I would see as mere opposition McGraw regards as a peril to his very being."

"Such men can be dangerous," said Mister Sonnheim.

"They can also be inspiring," said Mathewson. "When I learned that he was to manage the club I felt I'd found salvation. The thrill of big league ball fades swiftly with a last-place team. Having made the big leagues, having won twenty games, having pitched a no-hitter—" he tapped the ring against the tablecloth—"I had nothing further to prove to myself. What does it matter if you're not playing for the highest stakes? You've got to prove yourself in the pinch, where the pressure is the greatest. You've got to go through the fire!" His voice was rising, and heads were beginning to turn.

"Interesting," said Mister Sonnheim in a modulated tone; it

calmed the pitcher, and he relaxed as the footman poured coffee and offered sweets from a silver tray. "We come again to the warrior ethic," our host continued. "I found the Polo Grounds an attractive site, but I'd hardly choose it as a stage for Armageddon."

Mathewson laughed. "Nor would I, but McGraw could hardly imagine a better one, unless it was Columbia Park in Philadelphia or Comiskey's field out in Chicago. I strongly suspect that Armageddon will commence in the second week of October. No, no brandy, thank you. Have you a system for auction bidding, Mister Sonnheim? We'd best lay our strategy for the evening."

With the Giants assured of their league title, all New York pulled for Connie Mack's Philadelphians in the American League race. The reason was simple geography: the nearer city was accessible, while Chicago, close in the fight, was at the end of the civilized world, thought vulnerable to stampedes of cattle if not Indian attack. Most, too, assumed that McGraw would prefer to play the dour Mack, his opposite in every regard. Tall, aloof and aristocratic, Mack ran his club from the shadow of the dugout, dressed and mannered like a banker specializing in mortgage foreclosure. How different that Scotsman from the Irish McGraw, sharing his players' uniform and directing them from the coacher's box off third base!

But the Athletics appeared as eager to avoid the confrontation as New York ached for it. They slumped monumentally in September; their impeccable infielders threw away chance after chance; their best pitcher, the irrepressible Waddell, injured his shoulder in some clubhouse horseplay. Staggering home at season's end, they declared themselves champions although they'd played (and lost) four fewer games than the White Sox of the West.

Ban Johnson accepted their claim, calculating that the rivalry between the nation's two largest cities would bait the Giants into the contest, and he judged correctly. Brush of New York and Ben Shibe, the Philadelphia owner, went into a session that imitated the trappings of the Russo-Japanese negotiations of that same summer. Rumor reached its highest hour: Eli, claiming the inside dope, had it that the greatest argument raged over which club would gain the home date on Saturday. A coin flip settled the matter with a happy augury for the Giants. The Series would begin in Philadelphia on Monday, the ninth of October, and alternate daily between the cities until one club had four victories. Brush had pushed hard for a seven-game limit, Eli reported; the '03 Series had been a best-of-nine affair, but McGraw estimated that his two aces, Mathewson and McGinnity, were outmanned if not outmatched by Philadelphia's quartet of Plank, Bender, Waddell, and Coakley, and that the Giants' chances were better in a shorter set.

Eli copped every ticket that could be had, paying dearly for them; the scalpers demanded outrageous prices, and my brother dug deeply into his season's winnings to meet the cost. Our Philadelphia clients did the same at their end of the ninety-mile axis, but Columbia Park had not the capacity of the Polo Grounds and they came up shy for the opener. Eli, with a long face, told me that I could have a seat only at the expense of an insulted buyer and suggested that I apply to Mathewson for a pass of my own. This I would not do, nor allow Eli to make the request on my behalf. I said I'd be content to follow the action at Madison Square Garden, where play would be represented on a huge mechanical board. I thought it an acceptable sacrifice; perhaps McGraw would start with the Iron Man, as on Opening Day, and I'd see Mathewson on Tuesday at the Polo Grounds. I was wrong. Joe had slipped in the second half of the season, finally paying a toll for his unstinting exertions. He'd won eleven fewer games than Mathewson's thirty-two and trailed the collegian in every category save the number of appearances. On Monday afternoon, when I pushed through the excited swirl under the Garden colonnade and stepped into the immense arena, the board had the batteries posted: New York, Mathewson and Bresnahan; Philadelphia, Plank and Schreckengost.

I'd followed out-of-town contests in a similar manner for years. There wasn't a city tavern worth the name that didn't boast a ticker and a blackboard, and a lad in knickers to chalk the play. In comparison, though, the Garden operation was Gargantua to an organ-grinder's monkey. Their board was twenty yards square, the letters a foot high, and the painted diamond seemed large enough to accommodate an actual game. With admirable dexterity the boardmen maneuvered small cutout figures over the face of the diamond with long bamboo poles, while a barker cried out the progress of the game through a huge megaphone. We had the play not five seconds after it occurred ninety miles away. Our numbers on Twenty-Fourth Street nearly matched the attendance at Columbia Park, and the din, enclosed within the Garden and amplified by its echoes, was greater.

There were howls of anger when Plank's first pitch hit Bresnahan's leg, groans when the Giant was forced at second, wild excitement when McGann doubled. The thousands hollered encouragement as if the team could hear the cries, and cursed equally when a strikeout ended the Giants' first without a score. There was a great ovation at the barker's cry of "Mathewson on the mound." All could envision the blond master hitching his belt, flexing his right arm, and bending in to face the foe. They could not know, as I did, what measure of urgency and event the pitcher brought to the moment, nor could they feel as intimate with the cardboard cutout in the middle of the painted diamond. I saw him there as clearly as I had in St. Louis four years before, and his success was no less important to me now as then. Once more my focus narrowed until I seemed to live within him, to bend and sway and drive with him and to finish with my hands at the ready to field the ball.

My feeling did not evaporate with the first out or the first inning, or the first stretch through the Philadelphia line-up. In truth it lasted the week, in greater or lesser degree; it was a waking dream. Mathewson held the Athletics to four hits that Monday, walked none, struck out three; he singled in the fifth to begin a two-run rally and sacrificed in the ninth to build another run; the final score was three to nothing. On Tuesday I was at the Polo Grounds, electrified by the sight of the Giants' new black uniforms especially tailored for the event; a client, a Philadelphia partisan, joked that Eli surely had a counterpart in the garment trade. In the black togs McGinnity ever more resembled a mechanical contraption, but he lost to Chief Bender, three to nothing; Giant errors allowed the runs, and we were shocked while our guests gloated. Civility died that week. On Wednesday morning I departed New York under skies as black as the new Giant suits, and found the weather no better three hours later in Philadelphia; I spent the day in the lobby of the Bellevue, first participating in the scramble for rooms, then observing it after Eli's bribe had secured us a space. On Thursday, dizzy from the fumes that wafted over Columbia Park out of nearby Brewerytown, I watched Mathewson square off against Coakley. Again Bresnahan was hit by a pitch in the first inning; seven frames later, secure in a seven-run lead, Mathewson revenged his catcher by knocking Coakley down, and wrapped up another shut-out with two ninth-inning strikeouts. On Friday, back in New York, McGinnity showed his best stuff and outlasted Plank in a one-run drama. Joe's dedicated workmanship contrasted with Mathewson's jovial rooting on the bench; in the box seats there was knotting tension, then ultimate ecstasy.

The clubs stayed over that night, for Brush would not be denied his Saturday crowd. In the morning I attended services with Edith in her father's pew at Temple Emanuel across from the Public Library on Fifth Avenue. It was our practice to worship with her family on alternate sabbaths, but even the less traditional Reformed service observing the holy days of the New Year seemed interminable. Afterward I saw stark disapproval in Mister Sonnheim's eyes when I turned down the invitation to his home. Instead I handed my tallith and yarmulke to Edith and boarded a northbound trolley. The Stars and Stripes waved from a hundred flagpoles above the Polo Grounds, that secular house of worship. The queue for general admission tickets stretched half a mile. If the Giants were to win the championship at home, this must be the day; under the agreement reached after the rain-out, all further games would be in Philadelphia.

The early line listed Red Ames as the New York pitcher, and the odds favored the visitors, with Bender. At twenty minutes to three, after fielding and batting drills, the Chief walked to the practice slab on the visitors' side of home, with Powers to catch for him; from the opposite bench emerged Bresnahan, and behind him not Ames, but Mathewson.

At the sight of him there was a rush to find the bookmakers. All odds changed while Mathewson's chances with a day's rest were reckoned against Bender's, who'd last pitched on Tuesday.

New York's backers took heart from Mathewson's own demeanor. He seemed as loose and easy for his fortieth start of the year as for his first; he joked with the players and the front row spectators, and acknowledged an ovation by lifting not his own cap but McGinnity's, as the Iron Man stood beside him.

He struggled at the start. Yes, he held them scoreless, but not until the fourth inning did he put the side down in order. Bender more than matched him, and as yet another duel of arms unfolded the tension began to tell in the temper of the fractious fans and the intensity of the play on the field.

Bender slipped in the fifth. He walked a man to begin the inning and tried to tempt the next to bunt a high pitch, but he had not Mathewson's control and walked him as well. A sacrifice moved the runners over. Gilbert stood in, with Mathewson in the on-deck circle. He lifted a fly ball to left field; the runners held their bases as the ball descended and shot off as it touched leather. Hartsel's throw to the plate was off line, and catcher Powers moved to his right to snare it, backhanded; he had no chance for the runner speeding home, and smartly he whipped the ball to third, where Lave Cross slapped a tag on the surprised Giant who hadn't bothered to slide. Did the out come before our man touched home? It was umpire Sheridan's call, and McGraw was on top of him, hollering and gesticulating. Both men disappeared in the circle of players that crowded about; soon Connie Mack was alone in his dugout, barred from the field by his civilian dress. He stood on the top step, his hands choking a rolled-up program, and as players began to jump, the Giants in elation, the Athletics in dismay, he smacked the program against the dugout roof and threw it to the ground. A Giant run was posted on the scoreboard.

"Can he make it stand up, sport?"

"He hasn't lost a lead all year," I said.

"It can't go on," said a client. "He can't shut us out forever, not on one day's rest."

"He hasn't lost a lead all year," I repeated, and turned my attention back to the diamond, watching Mathewson stride to the mound to begin the sixth. He made his own trouble then, muffing a ground ball, and the client muttered, "It's telling. It has to tell." He could hardly be heard over the chant of *Bristol Robotham*

Lord. No enemy crowd could resist the rhythmic appeal of Bris Lord's full name, and they recited it, clapping in time, while McGraw took it up on the Giant bench. The unfortunate rookie had a greater disadvantage than his name. I doubt he'd ever faced a fadeaway; Mathewson had fanned him five times in the week, and now he bounced the ball weakly to second, forcing the runner, and then seemed not to notice Mathewson's pitchout and Bresnahan's pick-off throw. McGann slapped his glove against Lord's hip, and the youngster wandered off in the general direction of the Philadelphia dugout, miserable and lost.

Mathewson led off the Giant sixth, waving his cap as applause thundered through the grandstand. He lifted an easy fly to center field but ran all out just the same, and when he returned to the dugout he threw a varsity sweater over his shoulders, careful of his right arm, and stayed on his feet to pace the boards and shout encouragement to his mates. Taking the mound in the seventh he stopped at three warm-up pitches and waved the hitter to the plate. He threw two strikes, then a ball; he gloved Bresnahan's throw, took a breath, and climbed the hill once more. He bent and stretched and drove toward the plate, and the fastball shot through for his third strikeout. He threw strikes to Murphy, who grounded to second, and strikes to Monte Cross, who grounded to shortstop.

It had grown cold, and the October sun hung low over the bluff. In the dugout Mathewson chatted easily with McGinnity while Bender set down the Giant side. He shrugged off his sweater, set his shoulders, and climbed the steps to the field, his thin, elongated shadow gliding over the infield grass. He wasted nothing now. Powers saw only strikes, and lifted the third on a fly to left field where Mertes, shading his eyes with his right hand, gloved the ball with his left. Bender took two strikes and swung and missed the third. Hartsel pounded a chop to the right side; McGann moved after it and the batter raced Mathewson to first. The pitcher stretched for McGann's toss and won the sprint by half a step. He kept on running, describing a wide circle over the diamond; he was the last Giant to reach the dugout, and he went directly to the bat rack to take his stick.

Gilbert preceded him, and in the on-deck circle McGraw put

an arm over the infielder's shoulder and whispered in his ear. Under this instruction Gilbert wasted as much time as the umpire would permit, then took two strikes before fouling off a pitch. He examined his bat microscopically until he detected a fault, ambled past Mathewson to the dugout, and studied the bat rack as if it were an intricate curiosity on which he'd never laid eyes, testing half a dozen bats until he found one to his liking. He walked back to the plate as if the way were uphill, and never was stance taken with such exact care. In contrast, Bender's motion and throw seemed as antic as the shimmy of a nickelodeon harem dancer. Gilbert, down to the last strike, had to swing, and he lined the pitch to Lord in center field.

Four pitches, all wide, passed by Mathewson. He trotted down the basepath as the Giant batboy ran out with his sweater. I watched McGraw's apache dance in the coacher's box; somewhere in the ballet was an instruction for Bresnahan. Mathewson edged off first base, the infielders leaned forward, and Bender set, kicked, and threw. Bresnahan's bat swept into the curved flight of the pitch and connected with a sweet, satisfying snap. The ball shot out to left field; Hartsel turned his back and ran head down toward the ropes and the crowds they fenced. Ball and fielder fell against the barrier together and were lost in the milling excitement. I searched out the field umpire and spied him behind second base; his fist had two fingers extended. Bresnahan slowed to trot to second base, and Mathewson skipped toward third with the glee of a schoolboy at recess.

Now the kill: thirty thousand screamed for it, and while the Philadelphia loyalists in our box shrank and paled I screamed for it; even Eli, careless of our clients, screamed for it, and I loved him. Browne, at the bat, studied McGraw's signals, and Mathewson, a step away from his manager, shouted at the batter through cupped hands. Bender threw. Browne punched at the ball, and it skimmed over the infield grass, eluding Bender's reach; Murphy had no play but to first, and Mathewson dashed to the plate. We howled. Mathewson ran on to the dugout, where he threw himself into the joyous embrace of his teammates. The clamor enveloping the flats hardly diminished as Donlin went down on strikes and New York took the field for the ninth inning.

The pale light of the afternoon played over the dusty diamond, rutted by a season's play, and the autumn wind described spontaneous patterns in the bending grass of the outfield. The fielders, white against the sateen ebony of their uniforms, fidgeted with excitement, and the crowds, like excited ants, pushed against the restraining ropes in the outfield and jammed the aisles of the lower field boxes. Mathewson kneaded the stained baseball in his hands, and the unhappy Bris Lord took his place at the plate. Mathewson threw a strike, and another. Lord chased the next and tapped the ball back at the Giant pitcher, who collected it, examined it as the rookie raced for first, and threw to McGann for the first out.

He threw a strike to Davis, and a ball, and another strike; then Davis axed the ball into the hard dirt in front of home. It bounced high as Mathewson glided in beneath it; he snatched it with his bare hand and snapped a throw to McGann for the second out.

The ball streaked around the infield. Incredibly, for the place had no room for more noise, the tumult grew louder yet. Mathewson took Devlin's toss from third and stood atop the mound, rubbing the ball, savoring the moment. I thought that if he could he would order time to stand still, leaving him suspended for eternity in this pose, at this breath, but only gods and artists can stop time. Mathewson had to pitch, erasing this moment, bringing on the next: the ball skittering toward shortstop, the fans tumbling over the gates, Dahlen fielding the grounder and throwing it to first, McGann catching it, stepping on the bag, and tossing the ball high in the air. It caught the sun, dropped toward a thousand reaching hands, and disappeared.

The grandstand emptied. Wave upon wave of celebrants danced over the field, tearing at the uniforms of the champions of the world. I kissed Eli and ran, racing ahead of him down the ramp to the lower grandstand and along the walkway toward right field, jamming in among the thousands who were quivering and jerking toward the clubhouse like tiny filings to a magnet. At the extremity of the grandstand a low fence gave under the assault, and we spilled into the ever-expanding circle around the clubhouse.

We'd wait until dark for a sight of our heroes; we'd wait until dawn if we must. We cheered and chanted and sang and clapped for the better part of an hour, calling out to the players who peered through the high clubhouse windows. Finally two appeared on the veranda above us. One was Mathewson, the other was Bresnahan, and they carried a large yellow banner; smiling, each fastened a corner of the banner to the railing, and at a nod from the larger man they let it unfurl to reveal the hand-lettered legend:

THE GIANTS
WORLD'S CHAMPIONS, 1905

Thunder shook the air, and out of it emerged the chant of *Matty! Matty! Matty!* None cried the name so long and loud as I, and none watched the triumphant Mathewson through such a mist of unashamed tears.

September 23, 1908—at New York

										R	**H**	**E**
CHICAGO	0	0	0	0	1	0	0	0	0	1	5	2
NEW YORK	0	0	0	0	0	1	0	0	0	1	7	0

BATTERIES: Chicago, Pfeister & Kling;
New York, Mathewson & Bresnahan.
The game a tie, no decision awarded.

STANDINGS OF THE CLUBS

	W	**L**	**PCT.**	**GB**
New York	87	50	.635	—
Chicago	90	53	.629	—
Pittsburgh	89	54	.622	1
Philadelphia	74	64	.536	13½
Cincinnati	68	73	.482	21
Boston	60	82	.423	29½
Brooklyn	48	92	.343	40½
St. Louis	47	95	.331	42½

THREE

LATER, McGraw spoke from the clubhouse veranda. He was brief. "I appreciate the great victory as well as you," he said. "I thank you for your patronage and hope to see you all next spring." Next spring! The victory was but an hour old, the spring infinitely distant.

Had there been eight sons in our brood the duty of the seventh would have been clear, but Arthur was the last of us and as free as his two sisters of the obligations that had bound his elders. The rising tide of the family fortune could have borne him through any university, but he had not the temper for a formal education; he was an habitual truant, albeit one finally traced to the reading room of the New York Public Library. Alone of us all he was American born and accustomed to the spoiling riches of the New World. His every inclination was smothered by encouragement. At seven he drew a picture; the next day there magically appeared in his room an easel and a full set of oils; he never touched them. At ten he sang a song; he was instantly whisked to a professor of voice; he was soon cast out for trolling obscene ditties *a capella.* At twelve he wrote a poem; had he written a dozen they'd have been privately published; he never showed another. Poor boy! He had not the benefit of deprivation. He'd played ball in his early years, but not well; chess held a greater appeal, and for physical exercise he fenced and rode. The family called him a clever fellow, supported him generously, and despaired of him.

He was with us at the outset of the western swing of 1908. He'd toured Europe the year before and thought to complement that with a progress through his native land. We didn't expect that he'd take an interest in our affairs, and when he sat through a series of presentations that first morning in Philadelphia we marked it up to curiosity. At the ballpark that afternoon he showed little enthusiasm for the game: in late innings, the Phils and Giants tied, he frequently glanced at his watch. He was remarkable, though, in attention to his scorecard, tracking the game mechanically and recording it in minute detail with a mark for every pitch. By game's end the card looked like a medieval woodcut. He kept a second set of figures on a separate page: the credits and debits of Eli's wagering. So much money had changed hands over so many years that I'd long ago turned a blind eye to it, but now I watched my older brother through the eyes of the younger and saw that the old days of double-or-nothing for the sake of the clients were clearly gone. Eli sought no killing and never plunged, but he ended eighty dollars to the good.

Again the next morning Arthur lounged through our presentations, his left leg draped over the arm of an easy chair. At lunch, when I listed the city's museums and historical sites, he asked instead what local stores carried our line and wondered if he might visit one. In eight years I'd never felt a similar inclination. That afternoon, alone, he toured Wanamaker's, and in the evening we found him at a writing desk in the lobby of the Bellevue, scratching out indecipherable figures on the hotel stationery. At a breakfast meeting on Wednesday he ignored the assigned places and set himself beside a senior buyer. How many daily customers had his store? How many daily transactions in the jewelry department? What was the average dollar volume in a year? In January? In July? What amount was the average sale? Such answers as he received he found insufficient. He complained of this to us and begged our assistance in compiling accurate statistics. That afternoon he hustled us into a cab and directed it to Wanamaker's; he marched us to the center of the main shopping floor and extracted from his pocket three small circular counting devices. I knew their use at the ballpark, where club officials clicked off admissions at the gate. Arthur pressed one into Eli's hand and another upon me, lined us up back to back, and charged me to count the shoppers entering from the north and east, and Eli those from the west and south.

"Why don't we watch one door and multiply by four?" asked Eli.

"That wouldn't be valid," said Arthur. "We can't assume that an equal proportion of customers enters by each door."

"What if someone comes in, goes out, and comes back in? Do we count him twice?"

"Eli, this is serious business. Please don't joke about it."

"I beg your pardon!"

"I'm going up to the jewelry department. Start counting"—he checked his watch—"now!" We began; Arthur observed us for a minute, then footed it off to the elevator bank. Eli and I stood amid the bustle of a Wanamaker's Wednesday, clicking like a pair of amorous crickets.

"He didn't say how long he expects us to stay at this, did he?" Eli asked over his shoulder.

"No. When do they close?"

"Close! I'm not standing here the whole afternoon. The game starts at four, and Matty's pitching."

Time crept slowly; I glanced at my counter after ten minutes and read the startling figure of two hundred and seventeen. "What's your count?" I asked.

"Look at that clown whose wife is trying on hats," Eli answered. "She looks in the mirror while he checks the price tag."

"You're cheating."

"No, I'm praying. I'm praying no one I know passes by. 'What are you doing?' they'll say. 'I'm counting customers on my little counter,' I'll answer. 'Why are you doing that?' they'll ask. 'Because my kid brother told me to,' I'll say. 'If your kid brother asked you to jump off the Franklin Bridge, would you do that?' they'll ask. 'My kid brother would be waiting below in a rowboat, with a stopwatch,' I'll answer. 'Besides,' I'll say, 'I have an obligation to encourage his enthusiasm and initiative. I'm his older brother. I may be a damn fool, but I'm his older brother. Once I encouraged another brother of mine to take up jewelry design, and do you know where he is today? Right behind me, clicking his little counter. Our family teems with genius.' "

"It's good to see him enthusiastic about something," I said.

"Sport, the kid's problem isn't too little enthusiasm, it's too much, in too many directions. I know what's going on in his head. He's been with us for three days, and he wants to take over the business. He's sure we're doing it all wrong." Eli sighed. "I'll tell you something, sport. If he stays with it for three years, not just three days, I'll let him have it all. I'll retire to a life of leisure, and you can make me a gold-plated counter to click off the days. Is it fifteen minutes yet?"

"Just about."

"I'm going to find him. If he's in the jewelry department, okay, but if I find him at the luncheon counter nibbling on a sandwich or chatting up a waitress, I'll—well, I'll join him." With that he was off, but five minutes later he was back, a red rash of anger coloring his face. He snatched his counter from my hand and took his position behind me.

"It seems I embarrassed King Arthur," he muttered. "He was with the head buyer, dazzling him with the precise and scientific nature of his survey, when who should toddle up but yours truly, the lookout to the west and south. I had to make up some fool excuse for the crime of leaving my post. I suppose I'm to be shot at sunrise. Cripes, I'm mad."

"Chalk it up to enthusiasm and initiative," I said.

Eli whispered a curse. "We have King Arthur's kind permission to knock off at three-thirty. Precisely three-thirty. When we get back to the Bellevue, though, I'm going to have a chat with dear brother Arthur about enthusiasm and initiative."

"Why don't you let me do it?"

"Perhaps you'd better. Right now I'm angry enough to strangle him in precisely twelve seconds with precisely three feet of braided manila hemp."

"Count, Eli, count."

"I'm counting, dammit!"—which remark caused heads to turn. "Oh, by the bye," Eli said, "we're to write down the figures every ten minutes and make note—I'm quoting exactly—'make note of any extraordinary differential.'" Eli nudged me with his elbow. "Note this as well. At the moment I perceive no extraordinary differential between young brother Arthur and a horse's ass."

*

Mathewson beat the Phils that afternoon. Our downcast clients, home partisans all, were nonetheless eager to follow Eli to the Giant clubhouse and meet the pitcher. I was uncomfortable on such occasions and avoided them when I could; when I could not I behaved with reserve, as did Mathewson. I needed no effusive greeting; honor enough to see on his hand the ring I'd designed for him following the '05 World's Series. It was simple and bold, in the experimental style then emerging from German schools: three diamonds set flush on a thin band of platinum, without embellishment of engraving or filigree. To Eli it was a curiosity, far outside the run of our line and nothing like the old-fashioned pieces thick with swirls and curlicues that McGraw had commissioned for his champions. We could find no more than half a dozen in that clubhouse, for there was a new generation of Giants. They'd failed to repeat in '06, when half the club was laid low by diphtheria. Mathewson himself had succumbed to the fever, and huge crowds had welled about the hospital to read hourly bulletins on his condition; when he rose from his bed and toured the wards the *World* published an extra: MATTY WALKS! His lost weeks cast a shadow over the season and tainted Chicago's pennant. McGraw exacted a penalty in the contracts he sent out that winter. Donlin was so outraged by the reduced figure that he sat out the '07 season, and other veterans let their unhappiness show in their play. The club dropped to fourth place, Mathewson had but twenty-three victories, and McGinnity lost more often than he won. It made a shambles of Eli's betting system; no longer did a daily wager on the Giants guarantee a profit.

McGraw cleaned house. The club he brought north in the spring of '08 had only Bresnahan, Devlin, and a chastened Donlin from the regulars of '05, though much of the pitching staff survived. Veterans were imported—Tenney and Bridwell to prop up the infield, old Cy Seymour to chase down fly balls—while youth vied for the other spots. Bright-eyed, laughing Larry Doyle broke in at second base, and on the bench, eager for a chance, were a brace of anxious rookies: Fletcher, McCormick, Herzog, Merkle, Snodgrass. With these, McGraw expected, his next title would be won.

Such changes hardly mattered to the wider world that believed that the Giants yet reigned over baseball and that Mathewson, the perfect Mathewson of the three shutouts, was the only Giant. He was Matty to all, or "Big Six," for his size; a letter dropped into a mailbox in Omaha with no more address on the envelope than the numeral *6* was duly delivered to the Polo Grounds. As if he had not sufficient virtue the penny prints assigned to him the purity of Galahad, and his teammates were amused to read that he never cursed, never drank, never played at cards. His appeal changed the very nature of the ballpark crowd. College graduates taking their places in the city's banks and brokerages thought of him almost as a classmate, and on weekday afternoons when the market closed they boarded the new subway lines at Wall Street and rode directly to the Polo Grounds to watch him work. On Saturdays they brought their fairer friends; the club gallantly admitted the ladies at half price, and the atmosphere in the box seats was entirely transformed. The women fairly swooned at the sight of him—handsome as any matinee idol, graceful as a Russian dancer—and he did not fail to play to this circle. He practiced a grand entrance, striding down the right field foul line as if it were a red carpet laid for royalty. Capless, his blond hair shining in the sun, his smile dazzling, he'd laugh at the taunts of opposing players and avenge their insults with the perfection of his performance. There on the hill, a foot above the rest, he was the golden idol of the land.

Everywhere there was evidence of his unique standing. We traveled with the club on the next day's train to Pittsburgh, and through the length of the train anyone who bore a hint of athletic form was accosted and interrogated:

"Are you a ballplayer?"

"Yes."

"You're not Mathewson?"

"No."

"Where is he?"

"I don't know."

"Are you sure? I just want to shake his hand."

"Sorry."

"Are you a ballplayer?"

"I'm afraid not."

"Have you seen Christy Mathewson?"

"No, I haven't."

"I just want to shake his hand."

"Sorry."

"Are you a ballplayer?"

"Yes."

"Do you know where Matty is? I just want to shake his hand."

"Sorry."

Mathewson was closeted with McGraw in a compartment near our own. Though it was a mere day trip of seven hours Eli had booked two compartments. Into the first went our sample cases, which must be attended by either Arthur or myself; in the second Eli entertained the Giants, who—save for Mathewson and McGraw—held only coach tickets. There Mike Donlin breakfasted on a bottle of Irish whiskey; there Cy Seymour broke out a deck of cards for McGinnity and Hooks Wiltse. I ate alone in the dining car, then relieved Arthur's watch on our goods so that he could take his meal. He came back with a copy of the *Saturday Evening Post,* removed his jacket and placed it carefully on a hanger, flopped down onto the seat opposite me, crossed his feet, and opened the magazine. I'd waited up for him the night before, to no avail; he'd returned at midnight from places unknown, and there'd been no opportunity to broach the matters on my mind. Here was the moment.

"Done with your figures, are you?" I asked.

"I've just begun, actually. There isn't much to be learned from a single store in a single city. It's too small a sampling. I intend to go on with this through St. Louis and perhaps beyond, when I'm on my own out west."

"I'm surprised at your interest."

"I know you are." He turned a page and read. We heard a shout from the adjacent compartment, and Arthur looked up and grimaced.

"Another hand won," I said.

"Or lost. The betting he does! I had no idea."

"It's not so much, really."

"Not so much! A new automobile four years running!"

"He'd have that if he never bet a cent. That's Eli's style, and he earns enough to pay for it."

"He may take enough to pay for it, but I don't know that he earns it. Do you have any idea what the buyers say about him when he's not about?"

"It hardly matters, does it, so long as they give him their orders."

"I don't believe that Eli has sold a piece of our jewelry in years. As best I can tell the whole business is subsisting on your designs and the reputation of the New York Giants."

"I'm sure you're wrong about the first, and as for the second——"

"It's an awfully precarious thing to depend upon." Arthur reached into a vest pocket, withdrew a silver case, opened it and took out a cigarette, tapped it against the casing, lit it, and drew deeply. Then he opened his magazine and began to read.

"I think I know how you feel," I said.

"There's no worse assumption."

"Let me speak my piece. The first time I took this swing with Eli I was quite put off by his behavior. It seemed so unlike my idea of what it ought to be. Of course I had a very narrow conception of the selling trade. I couldn't fathom what Eli was doing, and even more I feared I'd have to adopt his ways. But it turns out that that's not the way of things. Whatever his methods, you can't argue with the results."

"Indeed I might, but, to repeat myself, I've too small a sampling at the moment. If you think it contemptible to participate in my research I'll make other arrangements."

"Not at all! I'm happy to help you. I just think we ought to leave Eli to better things."

"Other things."

"If you will. The point remains that Eli deserves respect, not disdain or contempt."

Arthur put down the magazine. "Do you know, Jack, that your work stands out on the shelves?" he said. "It's far superior to the competition, and a good thing, too. Those displays are numbing. Something must be done about that. I'm disturbed that Eli's never so much as asked Wanamaker's for a separate display for our line."

"What chance is there of having one?"

"None at all, if it's never asked. What's this, now?"

The train was slowing, and over the hiss of the brakes we heard the thump of a bass drum and the glissando of a glockenspiel. I pulled the window down and stretched out for a look at the depot. Heads were poking out of other windows, obstructing my view, but the noise grew ever louder, the heroic rhythm of a marching band and the cry of voices shouting Mathewson's name. I heard Seymour and Wiltse, just ahead, describing the panorama to Eli:

"Looks like the high school band."

"There's women! Lots o' women!"

"Guy in the top hat must be the village squire."

"Big banner says 'Welcome Christy Mathewson' in great big letters, and then it's got 'and the New York Giants' down below, real tiny-like."

"Pretty good crowd for a town this size. 'Course, it's Matty's part of the country."

"I seen bigger in a lot smaller burgs."

"Hey, little girl! Yeah, you! You got a sister?"

We passed in review; every eye was turned to the window ahead. The band struck up another rouser, the train lurched forward, and I fell back into the compartment. Arthur, expressionless, watched through the window as the train picked up speed and left the station behind.

"I gather this is hardly unusual," he said.

"It's been the rule since aught-five," I said.

"I daresay he deserves it. He's an uncommon man for a ball-player."

"He's an uncommon man altogether."

Arthur shrugged. "That's a common opinion. He does carry about him a rather haughty aspect, don't you think? Something of the prince who condescends to walk among the people."

"Wherever did you get that idea?"

"Last night, at the Schuylkill Chess Club."

"You were there? You saw him there?"

"Unavoidably. He strode through the rooms like an icon on self-display."

"Did you speak with him?"

"Very little, beyond the attractions of the queen's pawn opening."

"You played with him?"

"No. He watched me play for a short time. The fuss people made! It rather"—he paused—"it reminded me a bit of you, to tell the truth."

"Thank you very much."

"Believe me, I meant no insult. On the contrary. The experience gave me a better appreciation of something I've heretofore considered rather queer. Finding that you're not alone in your estimation of the man puts you in a different light."

"More common, you mean."

"Less odd, certainly. And seeing that the adoration of the man is so general, I wonder how we might put it to our advantage. Look at this."

He thumbed through the magazine and displayed a page. It was an advertisement for a brand of shirt collars; a drawing of Mathewson's face filled half the page, and below it, over his signature, ran a flowery endorsement of the product. "If shirt collars, why not jewelry?" said Arthur. "Why not our entire line?"

"Don't be ridiculous."

"Ridiculous? Why shouldn't he be happy to do it?"

"I'd never ask him to do such a thing."

"Why not?"

"I'd sooner ask Edith to model in a store window!"

"But don't you see?" said Arthur. "The man's already modeling in a store window. We need only add to the display."

"I don't know anything about that. I do know that I'd never ask him for such a favor."

In his ceremonious fashion, Arthur lit another cigarette. "You know, Jack, I'm only trying to help. Eli uses the firm's connection with the team to such an extent that I wonder his cards don't read, 'by appointment to the Polo Grounds, jewelers to his royal highness John McGraw.' It seems a small leap from that to a more general announcement through the medium of magazine advertisement. I find exciting the prospect of marketing copies of the triple diamond you did for him. Such an original design!"

"I don't think you understand how offensive I find your suggestions."

"Offensive? No, I don't understand that. I certainly don't mean to offend. But aren't we in the business of selling jewelry?"

"Are you? Is that your intention? Or will you abandon it when you leave us next week and head west? What will catch your fancy then?"

Arthur nodded and drew on his cigarette. "You've a perfect right to say that, of course. To doubt my commitment. I'm rather surprised at myself, that the trade has caught my fancy. I never dreamed there were so many aspects to merchandising. It does engender thought. But they're merely thoughts. Though I make them as statements, essentially they're questions, and it's the essence of a question to be unanswerable."

"You sound like something out of a Cooper Union lecture."

"Vienna, actually. A compelling debate between two scholars of whom you've never heard." He turned a page of the magazine. "I doubt that even Christy Mathewson has heard of them."

"You don't impress me with these airs, Arthur. Scholarly citations don't wash with me. You're riding not fifty feet from one of the phenomenons of our age——"

"Phenomena."

"——and you haven't the wit to realize it. You want only to use it and cheapen it."

"A phenomenon of our age!" Arthur turned back the page and regarded the drawing in the advertisement. "To think of buying a shirt collar from a phenomenon of our age! Why, I might order half a dozen!"

From Pittsburgh the Giants doubled back to the east; we went on to Cincinnati and St. Louis, where Arthur departed our company for the hinterlands. It was well for Eli that he left us before our Chicago appointments; that city was Eli's favorite stop on the circuit, and his most perilous. He played for high stakes there, and bet with a will. It had become a joke between us; as we pulled into the Union Station he'd hand me our tickets forward and tell me not to trust him with our samples in that rugged and appealing city.

We had the luck to catch the Cubs playing the Pittsburgh nine at the West Side Grounds. Champions twice over in the National League, Chicago had squelched Detroit in four games in the World's Series of '07 after breaking the Giants' infant record for victories in a season. No finer defensive team existed. Frank Chance, their manager, had redefined play at first base, taking his position daringly distant from the bag and covering nearly as much territory as second baseman Johnny Evers. This Evers was regarded by no less an authority than Mathewson as the league's most dangerous hitter in a pinch, and his fiery will to win matched McGraw's own. You couldn't bunt on Battleaxe Steinfeldt at third; still less could you put the ball by him. At shortstop was Joe Tinker, the little Kansan with the lined, leathery face and weather eye of a midwestern farmer. Behind these were the mellifluous outfield trio of Sheckard, Slagle, and Schulte; there were daily arguments in the bleachers as to which was the best ballhawk in the league, let alone on the club. Johnny Kling's gutteral, often obscene, encouragements to his pitchers rang out from behind the plate to reach the ears of everyone in the park, and the staff he caught was deep and fine: Pfeister, Reulbach, Overall, Fraser, and above all Mordecai Peter Centennial Brown, Three-Fingered Brown, whose index finger was abbreviated at the knuckle and whose curve balls darted and jerked like trout on a hook. "I wonder if he gets anything special with that cut finger?" McGraw once mused. "If I'm ever convinced he does, I'll have a surgeon go to work on my entire pitching staff." No one quite believed he was joking.

These were the Cubs, hobbled by early season injury but now, under the kindly summer sun, pulling together to play their best ball. But Pittsburgh had Honus Wagner. He hit a home run that day, one of ten he hit that year, an astonishing total. I remember it exactly, for it was the first I'd ever seen belted out of the West Side Grounds. There was a man on third, and Wagner needed an authoritative fly ball to bring home the run. Reulbach, Chicago's pitcher, tried to cross him up with a soft curve. You might beat Honus Wagner, but you could never fool him; he timed his swing with the precision of a headsman, and bang! The ball soared skyward, and the voice of the crowd, dead silent in the immediate moment, rose with the ball, not as a cheer but as a gasp of awe. I don't know that they ever found the ball, but they scratched a mark in the pavement where it was seen to land, and after the game, and for days following, people gathered at the spot, looked back at the ballpark, and shook their heads in wonder.

Pittsburgh was the second choice in the race, after the Cubs; few gave the Giants a chance. But the club McGraw had built like a backyard hobbyist—a new gadget here, a reconstructed doodad there—held together. Among the rookies Doyle was excellent and McCormick effective, and Herzog was a wonder at the bat, though there was no place to hide him in the field. The old men, Seymour and Tenney, were productive, and Bridwell, at shortstop, was toughest when it meant the most. No one was lost to injury, and even the weather favored them; their thin pitching took extra rest when rain postponed a number of games at fortunate intervals. McGinnity was finished. Half the time he was listed to start he failed to toe the mark, and often he was relegated to the bullpen. Wiltse had a strong year, but Hooks was no Iron Man. The real difference was Mathewson. He never missed a turn, and between times McGraw would call on him in late innings to protect a precious lead. Every aspect of his game was at its sharpest edge: he'd strike out one man in five, walk one in thirty, allow half a dozen hits on an ordinary day, one or two on a great one. A third of his victories were shut-outs, as they had to be: many times one Giant run was the winning margin.

These three clubs gave us such a summer as the league had never seen, rising and falling and scraping against one another from June to September, never separated by more than a whisper. Whenever two of them met the third would skip over the ensuing carnage into first place. Giant fans packed the grandstand as quickly as the club could build new extensions; those who found no place at the park bought hourly editions of the newspapers to keep abreast. A speculator who in June offered Eli five dollars a ticket for our box seats upped the ante to ten dollars in July, twenty in August. That month, from Lake George, Edith asked a favor on her father's behalf: would I kindly report the scores in my daily letter? Inning by inning tallies were posted on the Exchange, bellowed by newsboys, scrawled in soap on storefront windows. I felt that all New York had become Coogan's Bluff, overlooking the Polo Grounds and reverberating with its echoes. Shows would succeed or flop, stocks would rise or fall, juries would acquit or convict in the temper of victory or defeat.

They turned into September with New York and Chicago tied, Pittsburgh half a game behind them. By Labor Day the Giants led by a game and a half. The next day Mathewson shut out Brooklyn in eleven innings; on Saturday he won his thirtieth victory, and tripled home two runs to boot. Two days later he won again, in relief; he came in with the score tied, one out, and two St. Louis runners aboard; he walked a man intentionally, struck out the next, and retired all the rest. On Friday, the eighteenth, while seven thousand New Yorkers rallied for William Jennings Bryan at City Hall Park, five times that number stormed the Polo Grounds for a climactic double-header against Pittsburgh. The Giants swept the pair; Mathewson threw another shut-out, his eleventh of the season. On that same afternoon Chicago lost in Philadelphia. Saturday morning's papers showed the Giants four and a half games ahead of the Cubs, five up on Pittsburgh, and the specs were selling tickets at face value outside the Polo Grounds. A ten-inning defeat, the club's first loss in two weeks, seemed hardly to matter. There was no ball on Sunday, the last day of summer, and I took Edith on the boat trip around Manhattan. Her pregnancy, hardly visible when she'd departed for Lake George, was now obvious, and it seemed that every nanny aboard had a guess at whether she was carrying a male or female heir. When we passed the Polo Grounds she asked how old the child might be before I took it to a ballgame, but my thoughts were on the present; in my imagination I saw the championship banner fluttering over the empty grandstand.

To the distracted city Arthur returned like a man from Mars. I'd thought or heard nothing about him in nearly three months, and when he pushed open the door to my office on Monday morning my first impression was that he'd grown noticeably taller. How odd, I thought, to spring up at his age, and then I saw his boots, chocolate brown and tooled in the Spanish manner and fitted with three-inch heels.

"You've joined the rodeo!" I said.

He laughed. "I bought them in Wyoming. Do you know, there's a town out there called Jackson's Hole? I saw it, the very thing. Not the town, the hole. I was told that Jackson is in there somewhere. Dead, of course. He drowned in it a couple of weeks after the Custer massacre. 'Massacree,' they pronounce it."

"Sit down, cowboy. Welcome home."

He sat, reached for his cigarette case, hesitated, and drew back his hand. "I hope I am welcome," he said.

"Of course you are."

"I've just seen Uncle Sid, and I wanted to see you and settle things between us."

"Done."

"Just like that? You're in a generous mood."

"Why not? We're three and a half games up with just two weeks to go."

"Ah, of course. I find the entire municipality at its wit's end over the fortunes of its ball team. Truly a phenomenon of our age."

My smile evaporated. "Arthur, if you want things settled between us it might be best to retire that phrase."

"But I agree, it is phenomenal! If you're to be angry at me when I agree with you, what might you do in the event we disagree?"

I turned to my work. "You've seen Uncle Sid, you said."

"Yes, I've asked him for a place in the firm."

"Really! What sort of place?"

"In sales. That is, to begin in sales. I'd want to become familiar with every aspect of the business, in time."

"I suppose you're going to tell me you've done some serious thinking about this."

"I have, you know. May I tell you what I've thought?"

"Arthur, you hardly need my blessing to join the firm. I've no objection. Uncle Sid must be delighted, though perhaps a bit dubious as to your worth. You'll convince him, no doubt. Have you talked with Eli?"

"I intend to see him immediately."

"Will you grant that he has somewhat more expertise than you?"

Now the cigarette case came out. "I believe that one can learn from the mistakes of others as well as from their example," he finally said.

"That's my very point, Arthur. You don't yet know enough to judge a mistake. Please, you must understand this. You know nothing about this business. What you have in abundance is the capacity to learn. I beg you to take your time. Don't try to turn things upside down at a stroke." I put out my hand. "I wish you success, I really do."

"Thank you," he said, and took my handshake.

"Now go and see Eli, and good luck."

"You won't regret this."

"Arthur, I'm not hiring you. The fact is that I'll have very little to do with you. I can't teach you anything about design. It's just something I do, I'm no instructor."

"May I tell you one more thing?" he said, standing at the door. "I'll be very proud to represent your work. I hope I can do it justice."

I smiled. "That won't be very hard at all."

Arthur stepped out into the hallway. "Where's Eli's room?" he asked.

"Next to Uncle Sid's corner office."

"Is that the way of things? The boss has the corner office?"

"Congratulations. You've learned your first truly important lesson in business."

I listened to the clipped tempo of his boots as he strode down the hall, then turned to my desk and began to scribble in my sketchpad. Everything took a diamond shape, nothing had worth. Designing the company line had become as routine as the ride to work on the "el." I threw the pad aside and took up the morning paper. There were four pages exclusively devoted to the pennant race, and other sections made reference to it: the editorial page featured a cartoon of a mandarin in robes, the caption reading, "The Chinaman Says: 'April showers bring September double-headers.' " The Giants faced five of them in the next ten days; the key series with Chicago would begin with one on the morrow. I turned again to the sports pages. A writer speculated on the Giants' pitching the rest of the way. Today it would be Mathewson; for tomorrow, and after, it was McGraw's decision and anyone's guess. *Kapinski,* I thought, imagining it in newsprint with a parenthetical L to indicate a lefthander. I leaned back in my chair and began to work to the Cubs' order. I jammed Sheckard with fastballs and threw an outside curve for a third strike. I loosened up Johnny Evers with a high one under his chin, and had him bailing out on the next pitch. I fooled Wildfire Schulte with a fadeaway. I had good stuff, putting down the first batter in every inning; I gave up a two-out single in the second and walked Frank Chance with two down in the fourth. I was ahead and cruising in the seventh, and Mathewson was warming in the bullpen in case I needed late-inning help, when an apprentice appeared at my door with a summons to Uncle Sid's office.

Eli and Arthur were with him. "Sport, do you know who buys our jewelry?" Eli asked as I entered.

"If it isn't people of refinement and taste I don't want to know," I said with a smile.

"Tell him, Arthur."

Arthur read from a notebook in his hand. "Between two and five PM on Wednesday, June the twenty-fourth, the jewelry department at Wanamaker's in Philadelphia recorded one hundred and fourteen separate purchases. Three were made by unescorted women. Three. Twenty-two were made by women in the company of other women. Thirty-seven were made by men, alone or with a friend, and fifty-two were made by couples. I thought the figure on couples might be skewed by the time of year. It was June, after all, and people get married in June."

"That's right!" said Uncle Sid from behind his desk. "See, he knows so much already!"

"However," Arthur continued, "I surveyed Cincinnati in very late June, the last two days in fact, and by the time we got to St. Louis it was July. Sure enough, there was a significant shift in favor of male purchasers, from thirty-two percent at Wanamaker's to forty-one percent in two stores in Pittsburgh, and finally fifty-three percent in Cincinnati, fifty-two percent in St. Louis, forty-nine percent in Kansas City, fifty-five percent in Omaha, sixty percent in Laramie, fifty-one percent in Denver—it goes on."

"I wouldn't doubt it," I said.

"Add those purchases made by couples and one finds that men, in whole or in part, account for well over three-quarters of all sales. And if we consider only men's jewelry the percentage mounts precipitously." He closed the book with a firm slap, and in the silence that followed I found the three of them looking at me.

"Yes?" I said.

"Jackie," said Eli, "what exactly is your objection to soliciting an endorsement from Mathewson?"

I closed my eyes. "It's out of the question," I said.

"Why?"

"It's simply out of the question," I repeated.

"That's no answer, sport."

"It's enough of an answer." I made for the door.

"Yakov! Sit down."

I obeyed Uncle Sid's instruction.

"Now, answer the question," he said. "What's wrong with Matson?"

"Nothing's wrong with Mathewson, Uncle Sid. Do you know who Mathewson is?"

"I know from Matson," he said. "I know who, and I know what, and I know how good. You don't walk in this town, you don't know from Matson. Wait!" He held up his hand to still my response. "Now, Eli tells me he can get Matson's name for our line. Avram tells me it will sell the line. If it's good business, I got a responsibility to do it!" He beat his open hand with his fist. "I got a responsibility to you. To your momma and poppa. To my sons. To poor Riva, may she rest. To the whole family! Now, can you tell me it's not good business?"

"We'll offer him two thousand dollars against a percentage of annual sales over and above the average of the past three years," said Eli. "He could make as much from this as he does pitching for the Giants."

"See? It's good for Matson, too."

I looked at Arthur, who returned my stare, and then at Eli. "Have you asked him already?" I said.

"Arthur gave me the idea half an hour ago, of course I haven't asked him," said Eli. "And I wouldn't ask him without checking with you first. God of light, you gave me such a going-over when I wanted to hire Warner——"

"This is different," I said.

"That's right, it is. That's why I want to know your objection. Arthur said you had one."

"One! I've no end of them."

"Let's hear them then, sport."

But I couldn't find words. "Eli, to use my—our connection with him in this way . . . It's wrong, Eli, it's just wrong."

"You make it sound like we're trying to take something from him," said Eli. "It's just the other way around. We want to give him two thousand dollars!"

"Minimum," said Arthur.

"What makes you think he'd do it?"

"Come on, sport!" Eli reached for a magazine on Uncle Sid's desk and opened it to the advertisement for Pearl Shirt Collars. "These people are probably strangers to him. We've known him since he was a rookie!"

"Why not McGraw?" I said. "He'd sign on in a minute."

"People don't like John McGraw, not outside of New York," said Arthur. "They'd run away from a line with his endorsement. But Mathewson! People love Christy Mathewson, he's talked of everywhere. You're right, Jack, he's a phenomenon of our age."

"Yakov, tell me why you don't like the idea. Tell me it's not good business."

"I can't tell you that, Uncle Sid."

"Then what?"

"Jack, you can create an entire Christy Mathewson line. Rings, cuff links, studs, clips——"

"A crown, sport! A crown for him to wear in our advertising!"

"Eli, that is the single worst idea I have ever heard in my life."

"All right, not a crown. But a whole line!"

"Yakov, I'm listening, but I don't hear nothing."

There was nothing I could say.

"Okay, it's decided. Eli, you talk to Matson."

"No!"

"Yakov, it's decided!"

"Let me do it," I said. "I'll write to him."

Arthur coughed. "I'm sure you could write an excellent letter, Jack, but perhaps someone else should read it before it's sent, on behalf of the company. To see that the offer is set forth in the proper way. That is to say, Jack, it isn't your side of the business. It's not design."

"I'll let you read the letter, Arthur."

"Oh, no, you misunderstand. I have no——"

"What are lawyers for?" said Eli. "Why did I send Carl through school? He can read it."

"We oughtn't send a letter until the season's over," I said. "Mathewson will have enough——"

"What season?"

"The baseball season, Uncle Sid. He'll have enough on his mind between now and the World's Series."

"But his price might go up considerably in that time," said Arthur.

"So will his value," said Eli.

"True, but with haste we might achieve the one without the expense of the other."

"I will not send the letter before the World's Series is over," I said. "Give me that, at least."

"Fine, sport, fine. Send it when you like."

"Before the end of October," said Arthur.

"It's decided," said Uncle Sid. "Now, you got anything else, *mein kinder?*"

Eli looked at his watch. "Arthur, why don't you come up to the Polo Grounds with me? We'll make it a celebration."

"Not today, Eli. I've calls to make after being away so long, and I want to spend some time with Mom and Pop."

"Good boy," said Uncle Sid.

"Tomorrow, then, or Wednesday," said Eli.

"How long does this baseball go on?" cried Uncle Sid. "When is this office gonna get back to work?"

"Not long now," said Eli. "It'll all be over in two more weeks."

Monday's loss to Pittsburgh—bad fielding permitted two third-inning runs, and Mathewson himself drove home the only Giant tally—coupled with a pair of Chicago victories diminished New York's lead and made Tuesday's double-header all the more vital. I met Eli at the park, and together we watched the teams cuff about in the early innings of the opener. The Giants went ahead by a run, but the Cubs chased Ames to the showers with a late charge. McGinnity gave up a key hit in relief, and in pursuit of it Mike Donlin went down as if shot. Clutching his ankle and twitching in pain, he was carried from the field. The final score was Chicago four, New York three, and when the second game began there was Three-Fingered Brown on the hill for the Cubs, painting the strike zone's edges. Tenney couldn't swing at all and had to yield to young Merkle at first base; McCormick, another youngster, subbed for Donlin in right. Bresnahan could hardly walk from the dugout to the catcher's box, but he stuck to it, guiding the rookie Crandall through the Cub order like a harsh schoolmaster. They made it to the seventh just a run behind, but McGinnity failed again in relief and Brown got his second wind. It ended three to one, and the Giants' margin in the standings, so imposing just four days before, shrank to the finest arithmetic edge. There was never such a silence as when the final out was made. The Giants trudged to the clubhouse, and the fans, who habitually broke onto the field to dance with their heroes in victory, allowed the defeated club to depart in peace.

Driving south, Eli guessed at the next day's pitching while I scribbled the standings as they'd run in the morning papers. "Our lead is exactly six percentage points over the Cubs," I reported.

"We're still on top," Eli said. "That kid Crandall looked okay today, didn't he? He was great against Brooklyn."

"I'd be great against Brooklyn."

"I bet you would. Do you ever throw any more?"

"As a matter of fact I pitched a strong seven innings yesterday morning in my office, before you called me in to Uncle Sid." I looked at my notes, tore them up, and threw the confetti in the air.

"What do you think of Arthur?" said Eli.

"I'm just as happy not thinking of Arthur."

"Hey, sport, you're really somewhere on the moon these days. There's more to life than a pennant race."

"That's what I'm thinking."

"Then think about this. When Arthur left us in St. Louis in July I figured all right, that's his career in the jewelry business. Next thing I hear he'll have run off with the circus, or some fool thing. All through the summer I expected word that he'd started a pineapple plantation in Hawaii or opened a whorehouse in San Francisco or—what do they do in Los Angeles?"

"Where?"

"Then he charged into my office yesterday wearing those boots —did you see those boots?—and I thought he'd bought a cattle ranch. After all, he can ride, can't he? But instead he looks me in the eye and says, 'Eli, who do you think buys our jewelry?' And he whips out his little notebook and starts reading these figures about customers in jewelry departments, and not just in Philly or Pittsburgh or St. Loo but Denver and Salt Lake and San Francisco. He's reading about eleven AM and three PM and all the rest, and I'm not listening to a word, or rather a number—the kid is all numbers, isn't he?—but I'm thinking, God of light, he stayed with it! All summer, when he might have been drilling for oil or studying earthquakes or goodness knows what—I mean, can you remember a time when he wasn't into some fool thing?—but all summer he's been visiting stores and taking these numbers, putting them together and figuring them out. I say, well, good for him! It looks like he's really got his ginger going in one direction now. I think that's terrific. So does Uncle Sid, and so do the folks. Some of his ideas aren't bad at all, like using a real man's man in our advertising. That's a dandy idea! After all, as he says, it's the men who buy our line."

"Sixty percent, in Laramie."

"First he suggested Jim Jeffries or Tommy Burns, if the nigger doesn't knock him out in Australia. But I said, why not Matty? He said you had a problem with that, so I took him in to Uncle Sid and talked it over, and then we called you in. I'm glad we got it all straightened out." Eli laughed. "Of course, some of his ideas should be kept under his hat for a while, I wouldn't let them loose on Uncle Sid just yet. He wants to change the name of the firm."

"To what? Arthur Kapinski and Company?"

"Kapp. He'll take the name on his cards. He's got a point, you know. You have to admit it sounds kind of funny. 'The Christy Mathewson line, from Pincus Jewelers.' "

"What does Arthur suggest?"

Eli turned the car into Central Park. "First thing he suggested was the National Jewelry Company. I had to break it to him that Yussel Toski's been working that one for twenty years. I thought of Giant Jewelers, but Arthur said it sounded like we made rings for elephants. He said Kapp Jewelers, but I know that Uncle Sid would die before he'd put any part of Pop's name on his company. Think about it, see what you can come up with. But the important thing is that Arthur really wants to put himself into the job. He's proved that much, and I think it's terrific. I can even learn to live with him on the road. But separate rooms, if you please. Not like the old days, right, sport? Now, buck up. We'll take the Cubs tomorrow, you wait and see."

The dawn brought a crystal autumn day; an augury, I hoped, and a positive one. I was clutching at anything to lighten my mood, and in that humor anything might darken it. Arthur was in my office.

"What are the Christmas specialties?" he asked as I worked my way around him to perform my morning routine: the water put to boil, the pencils sharpened and the styluses set out, the shades opened to allow the northern light.

"The Christmas specialties," I said. "We work well in advance of the season, you know. I have the full line designed before we leave on the swing. Cousin Ruby breaks them down for manufacture while we're away, and Eli gives him the numbers when we get back. Now—would you get out of the way, please?—now, the specialties are half a dozen or so new ideas that come out of the trip. I talk with the buyers, and if there's something they'd like beyond what we're showing and if it's not too complex I draw it when we're back and give it to Ruby. Why do you ask?"

"Ruby says that if you don't give him the Christmas specialties by Friday he won't be able to ship them in time."

"Good God, that's right! Where are the sketches? Arthur, would you move, please? Here they are. Oh, Lord, I suppose I can do finishes by Friday. Friday, you say?"

"According to Ruby."

"I'd better see him. Have you been down to the shop?"

"All of yesterday."

"So now you know that end of the business."

"Jack, I have the impression that you're less than delighted to have me with the firm."

I looked at him. "Arthur, you're ridiculously transparent. Some people know how to play both ends against the middle like a true card sharp, but you're pathetic."

Arthur laughed and lit a cigarette. "I'm sorry, but you must realize I'm very new at it. I'll try to improve."

"Stop it."

"I am sorry, truly I am, if you feel I'm stepping on your toes. I'm doing my best to avoid that. I think of you as the firm's most valuable asset."

"I'm sure you tell Eli the same thing."

"That, believe me, is very far from the truth."

"I suspect that you wouldn't know the truth if it stepped on your boots and ruined the shine."

"Oh, yes, I'd know it then! And the truth is that I have some appreciation of the artistic temperament, and for that reason I make allowance for your behavior."

"Don't call me an artist. It's a job."

"The triple diamond for Mathewson——"

"Was not part of the job. I'm going down to the shop."

"Until this afternoon, then. Eli's bringing some buyers to the Polo Grounds, and he wants me to meet them. Perhaps in the afterglow of a Mathewson victory your mood will be improved."

"I wouldn't count on it."

"On the victory? What's happened to your faith?"

"Shut up, Arthur." I walked by him and out the door.

We'd called the back room of our original store the "shop," and we'd stuck the name on the vast operational spaces that filled the four floors below our offices. There, dies were cut, pieces stamped and soldered, and gemstones, real and artificial, set in place by a roster of mechanics and journeyman jewelers. My cousin Ruby had charge of the place. It was he who broke my designs into constituent parts and devised the steps by which a single piece could be multiplied by ten thousand. Ruby was a pale copy of his father, so like him in stature that he never had to buy a suit or shirt or pair of shoes; Uncle Sid's fit him exactly. Missing in him was the old man's passion. Ruby was careful, thorough, uninspired. He couldn't imagine that I might give him a drawing impossible to translate into a workable process for manufacture, and he performed prodigies of microscopic engineering because he was too dull to fail at it.

"I hate to press you, Jackie," he said when I called at his windowless cubicle, "but we have no chance of shipping in October unless I have your specialty designs by Friday morning. We're really fighting the calendar."

"I'm sorry, Ruby, I've had the sketches since we came off the road. I just haven't put them together."

"Is something the matter? The baby, perhaps? Don't worry, becoming a father doesn't hurt a bit."

"No, Edith's fine, thanks."

"Do you think you can have drawings by Friday?"

"I suppose I'll have to. I'm sorry I've slowed you up."

"It's not too late. Oh, Jackie, about Arthur——"

"What about him?"

"Where does he fit in, exactly? We were tripping all over him yesterday. He's very bright, of course, but he has strange ideas."

"How so?"

"I was explaining how we have to retool the machinery according to your designs, and do you know what he said? He said, 'Why doesn't Jackie design things you've already cut dies for? Wouldn't that save a lot of money?' I said sure it would save money, but that wasn't the idea. He didn't seem to get that at all." Ruby shook his head. "Imagine fitting your designs to the machine, instead of fitting the machine to your designs. That would change everything, wouldn't it?"

"Well, he's nineteen, Ruby, and you always want to change everything when you're nineteen."

"I didn't."

"That's because you started working here when you were—what? Fourteen?"

"Thirteen, the day after my *bar mitzvah*. Twenty years ago next February."

"Happy anniversary."

"I suppose. Oh, Jackie, is there a ticket for me for tomorrow's game?"

"Eli keeps the box, Ruby, but I tell you what. If I don't have the specialties done by three o'clock tomorrow I'll stay and finish them, and you can go to the game."

"Take your time, then." Ruby smiled. "Someone said Mathewson's pitching today. Why don't they save him 'til tomorrow? Why send him out with one day's rest?"

"Someone's got to stop the slide, Ruby. They've lost four in a row. McGraw wants to win today. He'll worry about tomorrow tomorrow."

"That's the way machines break down."

"Tell it to McGraw."

"Tell him yourself. I've never met the man."

"I wish I hadn't."

I returned to my office, pasted my sketches to the wall, and prepared my drawing board, but with the stylus poised over the paper I found it impossible to draw the first line. To be a pitcher! I thought. A pitcher, standing at the axis of event, or a catcher with the God-view of the play all before him; to be a shortstop, lord of the infield, or a center fielder with unchallenged claim to all the territory one's speed and skill could command; to perform the spontaneous acrobatics of the third baseman or the practiced ballet of the man at second, or to run and throw with the absolute commitment of the outfielder! And to live in a world without grays, where all decisions were final: ball or strike, safe or out, the game won or lost beyond question or appeal.

I would not work at the family trade this day; there was a ballgame at the Polo Grounds, and Mathewson was pitching.

There were long lines at the general admission gates, though tickets wouldn't be sold for the better part of an hour. The way to the box seats was shuttered. I looked up to the Bluff, and remembered how Eli had half pushed and half carried me up the steep slope the first time we'd ever come to the park. The memory drew me up a well-worn path, and when I reached the crest of the Bluff I saw tiny figures climbing the stairs to the clubhouse. I remembered the time—not so long ago!—when the clubhouse had stood like a lonely lighthouse on an outfield shoal. Now the bleachers abutted it, and the grandstand itself, once an intimate crescent around home plate, was a double-decked horseshoe of phenomenal size. Its roof blocked the view of home plate and third base. I thought of our bets and bluffs with derbied Johnnie on the day of McGraw's debut; he and his Irish fellows might well be among the mob pushing at the grandstand gate. I squatted and fingered some pebbles, flipping them down the slope of the Bluff, and I heard Eli's voice, a younger, far happier voice than the one I knew now, explaining to me that the tall man in the middle of the diamond below was the very same Amos Rusie whose name I'd learned to read in the American-language newspapers.

A solitary uniformed player descended the clubhouse stairs and walked slowly over the grass. His spikes left a clean, straight track in the clay as he crossed the infield. He climbed the mound, taking no stance, standing there with his eyes cast down, searching for signs in the dust. It cheered and frightened me to watch him summon the strength to bear the weight of a city's hopes. How lonely that circle of clay, how exposed the man who stood there!

Suddenly a great shout arose, and the bleachers swarmed with color and noise. The fans, discovering their hero on the mound, hailed him, and Mathewson, his communion interrupted, acknowledged their cheers with a broad wave of his cap. This was the public image I'd seen a hundred times; this was Matty. More Giants came onto the field, the quick running over the outfield like roistering bucks while the lame and the weary trudged to the bench and sat there nursing their aches. Last to appear was McGraw, no longer the lean spider of his playing days, puffy in the face and broadening at the belt. I watched the players rehearse under his baton, and I thought of them and their actions as pieces of a great and intricate design. In isolation each of their skills signified nothing, like the separate parts that spilled out of the dumb machines at the shop. The game was the process that welded them into a meaningful form, and the pitcher was the gemstone.

I descended the Bluff and walked through the gate to the box seats, bought a scorecard from a boy whose youth astonished me, and studied the batting orders of the two clubs. One would sleep tonight in first place, the other would not sleep at all. At McGraw's great shout the Giants sprinted for the clubhouse, leaving their wounded far in their wake. As they disappeared the Chicago players began to emerge from the opposite clubhouse door. When the booing was at its height I heard a faint voice calling my name, and I turned to see Eli and Arthur guiding their guests along the aisle toward our box. For such a game only prime clients received an invitation; these two bought jewelry for the R.H. Macy store. With room for only four in the box, Arthur bid the buyers farewell and ambled off toward the bleachers. There was no pretense at talking business; the matter was, who's playing? Who's hurt? Of our troop Donlin and Bresnahan would give it a try, but the right side of our infield was gone: Herzog would play for Doyle, and Merkle for Tenney at first base. The Cubs? September had exacted its toll on their roster: Hofman for Slagle in center field, Haydon for Sheckard in left.

Done with practice now, the Cubs gathered in their dugout while the Giants marched down the foul line toward their bench. This time McGraw led the way, and Mathewson was the last man on the field, responding to the enormous ovation with his familiar extravagant smile. Matty, Big Six, the master of them all at the height of his powers, the nation's favorite—yet if he were certain to win there would be no drama, no tension, and tension there was, in every corner of the park, on every street of the city, and, yes, in Chicago as well, where they gathered at taverns and tickers to follow the play and root for their own. They would be taking the play in Pittsburgh, too, in Cincinnati and St. Louis and cities that had never seen these men save on cigarette cards and etched woodcuts; everywhere that wires ran the game would reach. And now it began.

Johnny Evers, that hard, pitiless man, beloved everywhere in Chicago save his own team's clubhouse, singled in the first inning, but Mathewson struck out Wildfire Schulte and escaped the threat. Pfeister, for the Cubs, had almost as much on the ball, but he couldn't find the plate in the early going. In the second inning he unleashed a terrifying fastball at McCormick; the youngster took it in the gut and fell into the dirt at home, writhing and gasping for breath. Yet in two minutes he was walking unsteadily to first base; there was no one to take his place. In the fourth it was Evers again, beating Mathewson's best fastball for another hit and from first base gesturing at McGraw in signs it took no expert to decode. Once more Mathewson pitched out of trouble; the teams were scoreless through four, and the tension ate at us all.

In Chicago's fifth it broke in a terrible moment. Tinker, batting with one out, reached for a fadeaway and put it on a low line to right-center field. Donlin, forgetting his pain, or perhaps anesthetized to it with an Irish remedy, circled the ball, the better to field it on the charge and come up throwing, but when he reached he missed it cleanly. It rolled on toward the fence, Cy Seymour chasing it, while Donlin fell to his knees and pounded the grass in anguish. Tinker charged for third, cut the base flying, and headed home. It was a near play at the plate, but Joe slid across with the game's first run.

It might be the only run, for Pfeister had found himself. He dispatched the Giants easily in the home fifth. Mathewson returned to strike out two and retire Evers at last. Herzog, whose bat was gold even if his glove was dross, led off the Giant sixth; he hit a sharp one-hopper to third, and Steinfeldt, who made his living with his hands, stopped it, scrambled for it, and threw hurriedly to Chance, or rather over Chance; Frank extended every inch of his huge frame, but the ball cleared his glove by a fraction and bounced on toward the box seats. Herzog held at second base while the home fans went wild with hope and relief.

Bresnahan, utterly professional, impervious to pain, bunted Herzog to third; Steinfeldt handled the play cleanly and was mocked with the crowd's applause. Now Donlin, and redemption: the ball skimming over second base, Herzog trotting home, the score tied.

In the pandemonium Pfeister took hold of himself and the inning. Donlin moved no farther; Seymour and Devlin went down on curve balls. Came the seventh inning and it was one to one; came the eighth and the score was the same, and then we were at the ninth, with demon Evers leading off for the Cubs. Here was the ball game, the pennant, the season: who could doubt it? And here was Mathewson's fastball, and Evers missing it; and Mathewson's curve, and Evers swinging through it; and another fastball tucked in under Johnny's fists, strike three. The park exploded with noise.

Wildfire Schulte twice struck out: he took a fastball for a strike, swung at another and missed, held off a fadeaway that just missed the plate (and had McGraw livid at Umpire O'Day's call), and made ready for the one-and-two pitch. I expected an inside fastball; instead it was a curve that surprised Schulte as badly as it would have done me, but only Wildfire had to suffer the strikeout. Two away.

Frank Chance, soul of the Cubs: he swung at the first pitch and popped it foul into the box seats. The fans scrambled about, and one of them snatched the ball as it rattled under the seats and flipped it back onto the field. The batboy ran it out to Mathewson, who turned his back to the plate and massaged the scuffed brown baseball in his huge hands. That ball, bright white ninety minutes before, indistinguishable from a thousand others, now bore its own unique markings: Mathewson's hands ran over the smudge that Donlin's hit had made, touched the seam frayed by Tinker's bat. The game's history was on that ball, more exactly than in any box score or the memory of any witness.

Mathewson took his sign and pitched, Chance swung, and the ball bounced easily to shortstop. Bridwell's throw was true. Three hands out, all out: the first and greatest of Doubleday's rules, three-quarters of a century old.

Seymour grounded out to begin the Giant ninth. Devlin now, hitless to this point against Pfeister's fine curve, and the pitcher threw another, high, and yet another—good, too good, a batting practice pitch, and Devlin knocked it cleanly into center field.

Young McCormick was certain to bunt, and indeed he tried, once, twice, both times foul. McGraw's strategy foundered on the rookie's unfinished skills. Swinging with two strikes, McCormick grounded the ball to second. The Cubs tried for two, but Devlin's hard slide upset Tinker at second, and McCormick was safe by a fraction at first. Chance argued the call with field umpire Emslie; for a moment Emslie was our hero. A good call, a great call. Good old Emslie.

Two outs now, and Merkle at the bat: Merkle, the baby of the club, nineteen years old yet, big as he was, seeming far younger. He'd passed the year as a late-inning body inserted now at first, now in the outfield; with Tenney injured, this day marked his first starting assignment. He was hitless in three tries, victim of the rookie's devil, the major league curve ball. His bat never moved as Pfeister took him to two strikes. The third pitch was outside, the fourth closer to the edge: Merkle stroked it, and it rose over Chance's glove at first, slicing for the foul line. Every Giant loyalist stretched and strained to push or pray the ball fair. It nicked the line, sending up a white puff of lime, and the crowd shrieked. McCormick tore around second and dug for third; Merkle took a wide turn at first, thought about stretching the hit, played it safe, and returned to the bag. There he stood, unable to force back a happy, relieved smile while the cheers thundered down.

Bridwell advanced to the plate. The back rows of the grandstand emptied; the box seats were inundated with swaying, shouting men. Pfeister set and pitched; Bridwell swung; Pfeister brought up his glove but grabbed only air. Emslie, just inside second base, whirled out of the ball's path, lost his balance, and fell. The ball kicked earth just where the infield dirt bordered the outfield grass and bounded gaily into center field. From third base McCormick dashed down the line; from the coacher's box alongside McGraw raced with him. The rookie took two long steps, jumped, and came down on home plate with both feet.

The ecstasy! Fans burst onto the field and swarmed about the victorious Giants, who fought their way toward the clubhouse amid swirls of dust and turf. I was standing on my seat, both fists raised to heaven, and Eli was pounding me on the shoulders and shouting at the top of his lungs. Our clients threw their bowlers into the air, lost them to the lower deck, grabbed others that showered down upon us, and flung those after their own. I could hardly see the surface of the field, so crowded it was with celebration. Had there ever been a more glorious victory?

I had my arms around Eli; together we jumped and danced. "Let's go to the clubhouse!" I shouted. "I've got to see him!"

"You bet! But wait until things calm down!"

I looked down at the field again. Behind second base, standing out in their uniforms, were a trio of players, two Cubs, one Giant: I recognized McGinnity. Doubtless the Iron Man had claimed the game ball, which was Mathewson's by right and custom, but the Cubs—Evers? Hofman? I thought so—were contesting his title to it. There was a short scuffle, and Umpire Emslie came between them; McGinnity broke loose and threw the ball high and far away. I watched its flight, instantly reminded of the finish of the World's Series when McGann had flung the ball into the sun. This one came down amid the crowd at the base of the bleachers. I felt a flash of anger at the Chicagoans. To fuss over a souvenir that indisputably belonged to the Giants! It was hardly sportsmanlike. Then I looked toward the clubhouse and saw Mathewson riding on the shoulders of the adoring mob, and I began to cheer again. Forget Bryan, forget Taft: Mathewson for President! We laughed. And Merkle for Vice President; the rookie deserved a share of the palm. So did Herzog, and Bridwell, and Donlin, and good old Devlin.

There were boos rising from the field. Evers was still at second base, holding a baseball and talking insistently to both umpires, Emslie and O'Day. The three men made an odd progress through the milling celebrants in the outfield, the umpires walking away, Evers running ahead and turning to confront them, a few moment's argument, and then the whole process repeated. A passing fan took a swipe at Evers' cap, and Evers swung at him with his glove as if brushing away a noisome fly. I lost sight of them in the throng that circled the clubhouse. Thousands were streaming through the gates and onto the street while hundreds remained on the field, unwilling to depart until they had wrung the last measure of rapture from the afternoon.

I'd never been eager to visit the Giant clubhouse, but now I longed to be there, alongside Mathewson. I'd seen his private devotions four hours before and felt that my very witness had added to his strength; to be with him now, to confess my faith and share his glory was all my yearning. At the top of the ramp to the lower deck I threw my arm over Eli's shoulder, and marching in step with him I began to sing:

Take me out to the ballgame!
Take me out to the crowd!
Buy me some peanuts and crackerjacks,
I don't care if I never get back!
For it's root, root, root for the Giants!
If they don't win it's a shame—
For it's ein! zwei! drei *strikes* unt raus
At the old ballgame!

We struck up a second chorus as around us people joined in the song, amused by the interpolations which were ours alone. There is nothing that can so delight the heart as being a boy again, and there is no better transport to that happy past than a victory in a boy's game.

"Stick close now," Eli cautioned the clients as we pushed into the crush near the clubhouse. "I'll see you in." We fought our way up the stairway to the heavy iron door; Eli hailed the uniformed policeman guarding the passage, and he beckoned us forward. Eli clapped him on the back as he admitted us.

The room was large, almost square, and dimly lit; six caged hanging lamps added little to the twilight that filtered through the high windows. To call it a locker room would miss the mark; there were no lockers, merely long stretches of wooden shelves with nails beaten into the wall beneath to serve as clothes hangers. Milking stools were scattered here and there. Well-wishers far outnumbered the ball players; most of the athletes were in the shower, and their laughter and rough insults were amplified by the tiles. In a tiny corner office a knot of reporters was gathered around a wooden desk, and behind it sat McGraw, his uniform shirt unbuttoned and his stockinged feet crossed on the desktop.

Eli and his guests stood by the doorway to hear the exchange, but I moved on to the edge of a circle around Merkle. The boy's baby face was shining with delight.

"What was the pitch, Fred?"

"Curve ball, I expected it. Might have been a little outside of the plate. Don't tell Mister McGraw."

The reporters laughed. "I think he'd forgive you," said one.

"He's fined guys for less," said Merkle.

"What did you think when he told you you were starting today, Fred?"

"I just thought, please, God, don't let me embarrass myself. I was ready, though. You hate to see guys get hurt, but let's face it, that's the way rookies break into the line-up. Hey, we rooks didn't do too bad, did we? Moose scored the winning run, didn't he? And Buck scored the first one."

"How's it feel to be a game ahead?"

"Terrific!"

"What did you think when Bridwell's single went through in the ninth?"

"What did I think? I thought, 'great, we win!' Then I saw the pack coming out of the stands and I thought I'd better get the hell out of there." He grinned awkwardly, the boy using the man's cussword. "So I lit off for the clubhouse. Almost didn't make it. That was quite a mob out there, wasn't it?"

I saw Mathewson emerge from the shower, a towel around his midriff and another over his shoulders. He walked to the trainer's table and dropped the towels to the floor. Four years had added breadth to his shoulders and chest, and the color of his skin was slightly more pale, pearl-white rather than pink. He lay down on his stomach and stretched his arms to the edge of the table. The trainer took a palmful of wintergreen and began to knead the muscles of the pitcher's back. The smell of the ointment stung my nostrils. I would not approach him at this moment; it seemed unduly familiar. I'd wait until he was standing.

A large man in rimless glasses and a western hat was not so inhibited. He walked up to the table, ruffled the pitcher's hair, and said, "Not a bad outing, Matty."

The pitcher looked up. "Hello, Hugh. Knight to queen's bishop five."

The man laughed. "Are you telling me you've been figuring your next move all through the game?"

"I studied the board for half an hour after you left last night. I'll finish you off tonight."

"So you say. Still, I've a piece to write before then. Any pearls of wisdom for the public prints?"

"You said it yourself. Not a bad outing."

"Quoting one's self is the mark of a mediocre writer."

"Sounds like something you've said before." Both men laughed. "Really, Hugh, the kids are the story of this game," Mathewson said. "Talk to Buck Herzog or Fred Merkle. Or doesn't Hugh Fullerton condescend to interview rookies?"

I knew the name. Fullerton was a rare sort on the sports beat, literate and knowledgeable, no mere owner's pet, standing above his fellows as Mathewson did among ballplayers, writing for the gentlemanly press: his home paper was the *Chicago Tribune,* and in New York the *Times* ran his columns. In '06 he'd become a World's Series sensation, the only man to predict the White Sox upset of the Cubs, and his forecast the following year had great effect on the bookmaker's odds.

"You've given me my story," he said. "'Lordly Mathewson deigns not to talk with sportswriter. Shunts newsman to teammates.' One good quote, Matty, and then I'll see the kids."

"One good quote. All right, try this: 'He that lives this day and comes safe home will stand a-tiptoe when this day is named, and rouse him at the name of'—what? Fred Merkle?"

"'And gentlemen in Chicago now abed shall think themselves accursed they were not here.' Well, that's the truth of it, at least. It was a hell of a game, Matty."

"Yes, it was. It was a very devil of a game. Quote enough for you?"

"It'll do until tonight. See you later."

Fullerton stepped away from the table; the masseur applied another handful of ointment to the pitcher's right arm, worked it into the flesh, brought his hands down to Mathewson's wrist and back up to the shoulder, and patted him on the back. "You're done, Blondie," he said. Mathewson rolled onto his feet, his back to me, and wrapped himself in a clean towel. I was about to step toward him when I heard the commotion at McGraw's office. Reporters were backing out of the cubicle, and then McGraw appeared, his hands gripping the frame of the doorway, his knuckles white.

"Get the hell out! Out! Everybody! Everybody but the players! Dan! Get these sons of bitches out of here! Everybody out!"

Suddenly there was a mass of people between Mathewson and myself, and then I felt the cop's hand on my shoulder. "Come on, get a move on," he said. "Everybody out!" I accepted the order, but many did not and confusion spread. I saw a clubhouse boy with Merkle, pointing him toward McGraw's office. Then I was out on the veranda, already jammed to bursting. More men were pushed out, and finally I saw Eli backing through the door, chest to chest with the cop.

"Come on, Kappy, move along."

"Hey, Dan, take it easy, I'm no second-stringer."

"I know, Kappy, I know, and I ain't ever given you no shit, so don't start with me, okay? Show's over. Move along."

"Sure, Dan. Listen, will you let me know what goes on if you hear anything?"

"Sure, Kappy, sure."

The crush was becoming dangerous. I reached out and grabbed Eli's wrist, pulling him to me.

"What's going on?" I said.

"I don't know! We were just coming out of Mac's office when little Eddie rushed in and whispered something in Mac's ear. He went white and red and purple all at once, and then he started yelling for everyone to leave."

"What could it be?"

"I tell you I don't know, Jackie! Now let me see what I can find out!"

The wildest guesses were being made at the matter: an anarchist bomb was in the clubhouse, a fire had broken out in the grandstand, Evers had been shot. Eli spied our clients across the veranda, signaled his helplessness in the crush, and waved a goodbye, which they returned. We tried to push to the stairway, and I had a hand on the bannister when the door to the locker room swung open and McGraw burst out, his face a fury, a baseball bat in his hands.

"O'Day!" he screamed. "O'Day, you son of a bitch!"

He bulled his way to the stairway, the clubhouse boy close at his heels, and as they passed Eli swung into their wake and pulled me along. McGraw stormed down the steps and fought a path around the corner of the clubhouse to a small wooden door situated under the veranda. He grabbed the knob and rattled it. It didn't give.

"O'Day! O'Day, you fucking black Irishman, open this god damned door!" There was no response. "O'Day, you bleeding cocksucker, open this door or I'll break it to pieces, and then I'll start on you! O'Day!" He lifted the bat, and the crowd backed off to give him room. "O'Day, you've got ten seconds to open this door!" McGraw took a stance; there was total silence. "Five seconds, O'Day!" Nothing moved. Then McGraw started swinging, underlining each cut with a curse: "Bastard!" Crack! "Asshole!" Crack! "Shitball!" Crack! "Motherfucker!" Crack! On the fifth swing the heavy end of the bat smashed against the door's metal frame and snapped, leaving the manager with a foot-long splinter in his grip. He threw it to the ground.

"Somebody get me another fucking bat!" he cried, and he began to kick at the door with his stockinged foot. Just as he aimed a blow the door opened, and there stood Umpire O'Day in full working dress, his arms folded across his chest, his black short-billed cap pulled down to his brow. He was a head taller than McGraw, but the manager pushed his face under the umpire's chin and thundered:

"You cross-eyed, shit-faced son of a bitch! What do you mean, he didn't touch second base!"

"Get your spittle off my sleeve, Muggsy. I mean just what I say. Merkle is out. He's forced out at second."

"Forced out! Forced out! He could have paddled a fucking canoe to second base, you asshole! The ball was in center field!"

"Evers got the ball and touched second. Evers forced him out."

"When? Half an hour later?"

"It doesn't matter if it was half a year later, Muggsy. The force is in effect until the runner touches second base, and no run scores when a force-out ends the inning. The game is tied."

McGraw made tight pawing gestures inches from the umpire's face. "You're crazy! You're crazy! McCormick scored! The game was over!"

"Not until Merkle touches second, Muggsy. He cut straight from the basepath to the clubhouse, never touched the bag. Give him a roadmap next time."

"I'll give him a medal, you worm! You slug! My boys have orders to get the hell off that field when the game's over! Or maybe you'd rather see them torn apart by those idiots coming over the fence!"

"When the game's over I don't care if they whip out their dongs and pee on home plate, but the game wasn't over." O'Day took a small, dog-eared black book from his pocket, opened it to a marked page, and read: " 'One run shall be scored every time a base runner shall legally touch home plate before three men are put out'——"

"That's what happened, toad! When McCormick scored, the game was over!"

" 'Provided, however, that a run shall not count on a play in which a third man be forced out.' Read it yourself, Muggsy, I'll help you with the hard words."

McGraw slapped the rulebook to the ground and kicked it away. "You want rules? It wasn't even your call! It's the field umpire's call, and Emslie was on his kraut ass getting out of the way of the ball. How the hell do you know he didn't touch second? Emslie, get out here, you yellow heinie! Tell me he didn't touch second!"

"Are you telling me he did, Muggsy?"

"Call me that again and you'll be shitting teeth for a week!"

"Did he touch second base, Mister Manager McGraw, sir?"

"The winning run came home. The game was over and won. The kid ran to the clubhouse like he's told to do, like he's done every game he's played. Don't take it out on the kid, O'Day. Sweet Jesus, don't take it out on the kid!"

"I'm touched by your appeal to my better nature, weasel, but my report goes to the league office tonight. Merkle's forced at second base, it's the third out of the inning, the run doesn't score, and the game is tied. If you want to file a protest that's your right. Take it up with the league president."

"Pulliam? That double-barreled asshole! I had him on his knees in 'aught-four, begging me to play the World's Series."

"Looks like it's your turn to kneel, Muggsy. Grease yourself up good."

"Blow it out your ass, O'Day!"

"Get lost, McGraw. Get off my front porch or I'll have you suspended for the season."

"I'll suspend you from a fucking flagpole, you gutless stewpot! I'll have your job! I'll have your ass! I'll have your black Irish heart for breakfast!"

O'Day stepped back and slammed the door in McGraw's face. The manager kicked the dirt, stepped back, shook his fist in the air, and screamed, "I'll get you for this, you bog-crawling Mick! Nobody fucks with John McGraw!"

The door opened. "Especially Mrs. McGraw, you randy rat-faced turd," said the umpire, and he slammed the door shut once more.

The only sound was McGraw's heaving breath. Then a voice in the crowd shouted:

"Don't worry, Mac, he's not getting out of there!"

"We'll get him, Mac!"

"We'll fix his wagon, and Emslie's too!"

"Let's get him!"

"String him up!"

"SHUT UP!" McGraw roared. There was silence again. McGraw put his hand on the clubhouse boy's shoulder.

"Eddie, go get a cop. Get two cops. Eddie!" The boy stopped in his tracks. McGraw picked up the heavy end of the broken bat and with its splintered edge scratched a half-circle in the ground in front of the wooden door.

"Eddie, tell them to bring a couple of shotguns. And tell them that if any son of a bitch crosses this line, they should blow his fucking head into the river. Now, go!" The clubhouse boy shot off around the corner of the building, the mob making way for him.

"You! You!" McGraw said. Two men stepped out into the circle drawn in the dirt, and McGraw handed one of them the thick end of the bat. "Hold this position until the cops get here."

"Yes, sir!" said one, while the other said, "Yes, Mister McGraw," picked up the broken bat handle, and tapped it against his open hand while he eyed the crowd dangerously.

"All right," said McGraw. "Now, go home, boys. It's not up to the umpire any more. It's in Pulliam's hands now, and don't worry, boys. I've got Harry Pulliam right here!" He slapped his hip pocket, turned, and walked deliberately in the path the clubhouse boy had broken. The mob began to shout his name. He marched up the stairway to the great door of the locker room. Over the cheers, I heard it slam shut behind the manager.

Eli's face was white. "Didn't touch second base! God of light, I don't believe it! Listen, people, you have the wrong idea! The hell with O'Day! You should be stringing up Fred Merkle!"

"Eli——"

"Didn't touch second base! God of light! Do you know what that idiot cost me?"

A shiver went through me. "No, Eli, what did he cost you?"

Eli's voice was tight. "Never mind that, sport. Just believe me, we're a lot better off with a win than a tie."

"Pulliam will decide——"

"Pulliam! Harry Pulliam's been after McGraw for years. I was standing right next to Pulliam when he handed McGraw his championship ring, and do you know what he said? He said, 'I'll get even with you, you little bastard, if it's the last thing I do.' Leave it to Pulliam and he'll forfeit the game to the Cubs!" He closed his eyes and put his hand over his face. "God of light! Fred Merkle! Jackie, come with me to the Ansonia. That's where we'll get the word."

I adored the aspect of the Ansonia Hotel. For two years I'd watched swarms of masons inventing every manner of decoration for its ledges, its balconies, its apses and domes until it looked like an utterly gigantic and unbelievably complex cuneiform. It fronted Broadway along the whole block between Seventy-Third and Seventy-Fourth Streets, the largest hotel in the city and at every point its most ornate, the apotheosis of the Beaux Arts style. I felt that it would stand when every other structure in the city I knew had crumbled; it was our mightiest celebration of affected energy. It was, moreover, a people's hotel; the Waldorf, the Fifth Avenue, the new Plaza by the Park disdained all but the rich, but the Ansonia welcomed everyone, and for this it had quickly become a favorite of the people's entertainers. Cohan, Foy, John Drew stayed at the Ansonia, Tod Sloan in racing season, and the whole fistic mob in winter. The thick walls of its rooms were proof against disturbance; Rudolf Friml could rehearse productions in his huge rooms on the hotel's uppermost floor while in the next apartment Lillian Russell could sleep in peace. There was never an hour when some great luminary could not be seen and approached there; as this was known the lobby and public rooms were forever crowded with schemers and promoters who, like the pigeons in the thousand nooks of the decorative front, roosted there and fouled the place with their leavings.

It was close to dark when Eli and I arrived. We took a table for dinner, but my brother never sat for more than five minutes at a stretch; he was forever up and about, greeting (it seemed to me) every third person who came into the place and whispering his guesses at the course of events. He learned nothing to ease his mind. The Giants certainly had not won the game, and might yet lose it; Chicago claimed, with unarguable veracity, that the home club had failed to clear the field for the tenth inning and therefore ought to forfeit. To Giant fans—and everyone in the hotel was a Giant fan—the whole matter was ridiculous; the game had ended when McCormick touched home with the winning run. But the rulebook had it otherwise: Merkle had not touched second base.

It was sad to watch Eli scamper from place to place in a vain attempt to gain the inside dope. I'd always imagined that he had easy access to the innermost circles of concern, but the odd customers with odder names who came to our table could only repeat the common and incredible rumors on every tongue: Frank Chance had been seen handing O'Day five hundred dollars cash! said Harry the Hat; the police had O'Day in "protective custody"! said Bobby Bottles; Pulliam had fired O'Day and declared the

Giants winners! said Sandpaper Sam; Pulliam had suspended McGraw and forfeited all remaining Giant games! said Nick the Greek. And Merkle had committed suicide! ("Good!" ran the popular reaction to the tale.) When a broad-shouldered, bull-necked rowdy passed our table Eli reached out and took his elbow.

"Hal! Have a seat. Were you at the game?"

"Hello, Kappy. No, I wasn't there, I've had enough baseball for the season. Did he touch second?"

"He might have been the only man at the Polo Grounds who didn't, there was such a mob on the field. Hal, this is my brother Jackie. Sport, meet Hal Chase."

So this was Chase! He played first base for the Highlanders, the forlorn American League club that performed in near secrecy on the western heights far uptown. The franchise belonged to Tammany men who'd sold their interest in the Giants to John Brush; Andrew Freedman was among them, and the shift in leagues had changed him not at all. Managers rose and fell like beer in a bucket. Chase had campaigned for the job in the press, and when Freedman passed him over Chase quit outright, charging Freedman with a breach of promise and a failure of intelligence. Freedman denied the former accusation in terms that supported the latter. While his club slid to the bottom of the standings Chase encamped at the Ansonia, specifying his complaints and outlining his demands, to anyone who would listen.

"How do you see it, Hal?" Eli said. "Won't the league office overrule O'Day? You can't decide a pennant race on a stupid technicality."

"There's a joker in the deck, Kappy," said Chase. "The same damn thing happened to the Cubs three weeks ago in Pittsburgh: bottom of the ninth, two out, men at first and third. Base hit won the game, man at first ran directly to the dugout. Evers pulled the same trick, running for the ball and touching second in the middle of the mob. Guess who the umpire was?"

"Oh, no."

"That's right, hunkerin' Hank O'Day. He let the run stand, gave Pittsburgh the game, because he hadn't been watching the runner and couldn't take Evers' word for it. I bet Hank had a sharp eye out for Merkle when that hit went through today. He's got the rulebook on his side."

"God of light!"

"And then there's something about interference, I hear. McGinnity copped the ball when Evers was screaming for it."

"He did," I said. "I saw that. Joe threw it practically into the bleachers."

"Interference pure and simple," said Chase. "You'll be lucky the Cubs don't win by forfeit."

Eli moaned. "That's what will happen. Pulliam's been aching to shoot down McGraw since 'aught-four."

"The chickens are coming home to roost, Kappy. Maybe I'm better off getting out of this crazy coop altogether. They pay well in the California leagues, and there's plenty of action on the side."

"How do you see it ending, Hal? The pennant race, I mean."

"Pitching wins pennants, Kappy. Brown, Pfeister, Reulbach, Overall——"

"The Giants have Mathewson," I said.

"Matty's one man, the Cubs have a quartet. I figure the odds at four to one."

Eli looked up. "For how much?"

"What?"

"What will you take at four to one?"

Chase smiled. "Not me, Kappy. I've nothing to do with that race. I'll give you plenty on where the Highlanders finish, though."

"No, thanks."

Chase stood up. "Good to see you, Kappy. Look me up if you ever get out to the coast, I promise I'll put you on the inside track. Oh, here's an interesting souvenir you might want to paste on your shithouse wall." He took a rolled newspaper from his side pocket and dropped it onto the table. It unfolded to reveal the insignia of the *American* and the banner headline: GIANTS WIN, 2–1.

It grew late. Wandering through the lobby I saw more evidence of Eli's standing; the years of favors done and favors asked had availed him little. He approached club officials and high rollers with the same eager expression that marked the low life when they came to him, and he received the same joking dismissal. The embarrassment I felt never seemed to faze him. Finally I told Eli that I ought to phone Edith. He handed me a coin, thought for a moment, then gave me another and said,

"Why don't you call Matty? He ought to know what's going on."

"Eli, I can't telephone Christy Mathewson!"

"Why not? Didn't you talk to him in the clubhouse today?"

"I never got the chance."

"So take it now. Ask him what's going on with the protest."

"I won't."

"Why not? Listen, sport, we have more than just a fan's interest in the outcome."

"That's not my concern."

"God of light, Jackie, don't go high-hat on me now! Call Mathewson and ask him what he knows!"

"Call him yourself, if you must."

"Well, of course I could call him myself, sport, but you wanted to be the one to write the letter about the endorsement, so I figured——"

"That's another thing. How would it look, my calling him in the middle of the night to say——"

"It's just ten o'clock."

"——that my brother has a big bet riding on the game, and what does he know about it? Could I write to him in three weeks and offer him a business deal? How would that look?"

"It would look like the big, wide world, Jackie, it'd look like 'you scratch my back, I'll scratch yours.' Two thousand dollars is a damn good price to pay for the answer to one question, even in the middle of the night. We've been scratching Matty's back for years. How many rings have you done for him? How many gifts? It's time we got a little something back."

"Something back? He's given you something back every time you've come into the clubhouse with another set of fat clients! He can't come out of the shower without running into you with a couple of rubes or would-be sports grinning like idiots and grabbing his hand. When has he ever let you down, or told you to get off? Something back! Eli, what's wrong with you?"

We stared at one another, surprised with ourselves and sensing the danger in another word. "I have to call Edith, or leave right now," I said. "Do you want me to stay?"

But Eli was looking past me. "Here comes the answer to it all, sport," he said. "Here comes McGraw."

Not that the little man could be seen in the crowd surrounding him; it was the shouting of his name, like a ringside mob urging on its favorite, that announced him. With him was the central character in the drama. Fred Merkle towered above the manager, his face gray and his eyes blood red. I was intensely curious to study the rookie, and as he and McGraw pressed through the lobby in a path broken by a coterie of companions we elbowed our way after them. The party pressed into the men's bar and took a table. For all the noise in the room there was a growing quiet at the center so that the words spoken there might be overheard and repeated back over the shoulders of the onlookers; soon everyone who cared to know had it that McGraw wanted a steak and a baked potato, and that Merkle had muttered he wasn't hungry. Merkle was on McGraw's right, and Harry Fabian, the groundskeeper, on his left; four others completed the party at the circular table, including the clubhouse cop, now in civilian dress. I heard McGraw remark that there was too much salt on the pretzels, and instantly Fabian had the bowl in front of him, rubbing the crystals off with his napkin. An acrobatic waiter managed to push through to the table with a pitcher of dark beer. McGraw and Merkle abstained, the others poured and drank. The table talk deliberately avoided the main concern; the whole effort apparently aimed to show that no concern existed. Instead McGraw praised the Ansonia's kitchens to Merkle, promising many future visits. Gloom was all over Merkle's face as he sat with his shoulders hunched, looking at the tablecloth in front of him or at McGraw. McGraw's own eyes were never still; he winked and waved in all directions, including ours, and Eli returned the greeting. McGraw leaned toward the cop, pointed at us, and said something. The cop nodded, got up, and edged around the table toward us. Eli pushed forward to meet him, but he reached for my hand and said, "Mac wants to see you."

"Me! Why?"

He didn't answer, but pulled me forward. Eli tried to follow, but the cop said, "Sorry, Kappy, just the kid."

"But, Dan——"

"That's what the man said." Eli sagged, and his eyes urged me to do I knew not what. The cop led me around the table and sat me down between McGraw and Fabian. I saw the '04 club ring on McGraw's right hand.

"Little Kappy! Good to see you. You know Harry Fabian? Mickey Welch? Charlie King? Fred, have you met Little Kappy?" Merkle nodded, but didn't offer his hand.

"Well," said McGraw, "were you at the park today for that great circus?"

"I was there."

"I'm glad I spotted you," said McGraw. "I thought of something you might do for us. I guess you're ready with this year's ring, if the bastards don't steal it away from us." He laughed while the comment was repeated in whispers through the bar.

"You know, Little Kappy, these youngsters on the club are a bunch of world-beaters. I've never known a team to have such a crop in one year: Doyle, Fletcher, McCormick, Crandall, Herzog, Snodgrass, and Fred, here. They've kept us in the race. Like today, who won it for us? Who took Jack Pfeister's curve ball, the best in the league, and put it into the opposite field in the ninth? And who played it safe on the turn, just as he should?" He patted Merkle's arm. "This boy's a smart ballplayer, and anyone who says he isn't doesn't know bull about baseball."

"That's right, Mac." Three at the table said it at once. Merkle never looked up. I tried to keep my eyes on McGraw but couldn't help glancing at the rookie. He looked like the last survivor of an earthquake; his hands trembled, his face was a vacant horror. I knew that McGraw was improvising for Merkle's sake, and wondered if Merkle knew it as well.

"No, we wouldn't be close without these kids, and I want to do something special for them. I want a ring, a class ring, if you get my drift, for the Class of 'Aught-Eight. I think something with seven diamonds on it, doesn't that sound right? Charge whatever you want. And I want them by the end of the season, so I can give them to the boys before we play the World's Series."

"I don't know that we can do that," I said. "That's just two weeks away, and it's the busiest time of the year. I'm behind on my work as it is."

"Oh, come on, son, this is for John McGraw and the New York Giants. Drop everything else."

"I really don't think——"

"I want those rings on the last day of the season, son." The order might be a mere gesture, but McGraw's whole force was behind his words; I felt his power, and wondered how O'Day had stood up to him.

"I wouldn't wear it," said Merkle.

"What?"

"I said I wouldn't wear such a ring."

"Don't be ridiculous. You'll be proud to wear it." McGraw took hold of Merkle's wrist. "Better make Fred's extra large. Look at those hands! Look at that body! Give me a body like that and I'd have been twice the player I was."

The waiter put the steaks on the table, reached into his pocket, and handed McGraw a folded note.

"Eat hardy, Fred," the manager said. "You've got a ballgame tomorrow. You're my first baseman from—Jesus and all the saints!"

McGraw threw the note onto the table, jumped up, and plunged into the crowd, his napkin flapping over his shirtfront. His companions took off after him with equal alacrity, save for Merkle, who followed dumbly. The excited manager and his claque fought through the lobby and out onto Broadway, leaving seven heavily laden plates for seven grateful freeloaders to feast upon. For myself, I picked up the note and read: JOHN BRUSH COLLAPSED AT LAMBS CLUB.

Eli was at my shoulder. "God of light! Come on, sport, we've got to get over there!"

"Why?"

"Why! The president of the club drops dead, for all we know, and you ask why!"

"I'm tired, Eli. I'm going home."

"What did McGraw say? Where does the protest stand?"

"I don't know. I don't care. I'm going home."

"He didn't say anything?"

"Yes, Eli, he did. He said he wants seven rings made up for the rookies on the club, and he wants them in two weeks, and, Eli, I don't know how he's going to get them, because we're backed up at the shop already and besides, I'd no sooner design such a ring than Fred Merkle would wear one. I'm not about to celebrate this day. Go to the Lambs Club if you must, Eli, but I'm going home."

The next morning's papers were on sale at the kiosk at Broadway and Seventy-Second Street. I bought the *Times* and opened it to the report of the game.

Censurable stupidity on the part of player Merkle,

the story began,

> in yesterday's game at the Polo Grounds between the Giants and Chicago placed the New York team's chances of winning the pennant in jeopardy. His unusual conduct in the final inning of a great game perhaps deprived New York of a victory that would have been unquestionable had he not committed a breach in baseball play that resulted in Umpire O'Day's declaring the game a tie.

The rookie's crucifixion had begun. The fury of the season now lay upon the head of a nineteen-year-old Wisconsin farmboy less concerned with the technicalities of play than with self-preservation in the face of a mob. I'd always held that over the course of a season, with all its errors and accidents, the better team prevailed and the better players defeated the odds, that the game, at bottom, was just. Now I knew that the field of play was not exempt from life's injustices—a lesson nowhere heard in after-dinner speeches.

The lights were on in the parlor of our Nineteenth Street brownstone; Edith was entertaining a visitor. Arthur sat in a high wing-back chair, sipping tea in correct, polite posture. At seeing him my first thought was of his age—nineteen, like Merkle.

Edith made her excuses and left us, and Arthur poured himself another cup of tea.

"So you're taking a second floor," he said.

"And an option to buy the whole house."

"You're tidy with your money."

"Did you come to read my accounts?"

"No, I did not. I came bearing a gift."

"How kind."

"There it is again. You've built up a wall against me, Jack, but I refuse to let it stand between us."

"What can you expect, Arthur? You're a runaway motor. No matter what I say you do the opposite, generally five minutes later, and you're determined to be at odds with Eli. It's a family business, Arthur, for all its growing size, and you have to get along with the family."

"Indeed I do," he said, "and at birthdays, anniversaries, weddings, and funerals I intend to be the very model of decorum. At the office, however . . . do you recall Uncle Sid's comment on the Mathewson endorsement? 'If it's good for business I got a responsibility to do it.' " He aped the accent. " 'A responsibility to you, and you, and you, and your momma and poppa, and poor Riva, may she rest.' Jack, do you have any notion what Mathewson can do for the line?"

"That doesn't matter to me."

"It matters to me. It matters precisely because it is a family business. Why do you think I came along on the swing last summer? I could have found better company for travel, and heaven knows I care not at all if Pittsburgh defeats Cincinnati or St. Louis loses to Philadelphia. It's the business that interests me, and that's what I came to learn. I did learn two things, especially. The first is that we offer an outstanding line compared to the competition. That's to your credit, and Ruby's."

"I'm glad you appreciate Ruby."

"Indeed I do. Next to you he's the most important man in the company."

"Arthur, this flattery——"

"It is not flattery. I rate him below you because there are other men who could do Ruby's job as well as he, but your talent is unique. That's the very value of the artist. His work cannot be duplicated."

"Ruby's machines duplicate it ten thousand times over."

"You insist on misinterpreting what I say. Very well, let me tell you the second lesson I learned, which cannot be misinterpreted. Eli is an impossible gambler who ought not to be trusted with a position of responsibility."

"That's not so!"

"Indeed it is. It might not have been the case years ago, but it was clearly evident to me this summer. Jack, these aren't friendly wagers he's placing. You may not realize it—Eli himself doesn't realize it, I suspect—but the clients do, and they don't like it. Not at all. They order our line in spite of Eli, and absent Eli they'd order more. He's costing us business, and that means he's harming all of us, everyone in Uncle Sid's litany. He's harming you, harming Edith, harming your unborn child! I cannot let a vital part of the family business remain in his hands."

"I don't think it's the lost business that upsets you, Arthur. You're embarrassed to have him as a brother. You tolerate me because you have it in your head that I'm an artist, and you think that entitles me to—what? Eccentricity? But you don't allow Eli to be Eli."

"He may be whomsoever he wishes. He may not do so and retain a position of responsibility in the firm. I cannot fathom how one can trust a man who will lay stakes on an enterprise so fraught with accident as a baseball game. Look at this."

Arthur took a scuffed brown baseball from his pocket and held it up to my eyes. "I was all over the park today, Jack, the grandstand, the bleachers, everywhere. When the last run was scored and the whole crowd came onto the field I came with them. I could hardly help it, I was carried along with the mob. It's an extraordinary feeling to be on a major league field, isn't it?"

"I wouldn't know. I've never set foot on one."

"Haven't you? Not even to leave the park, as so many do?"

"Never. I never shall. It's a player's place, not mine. I never earned my way, and I'd feel it cheating to steal onto the field in the midst of a mob."

"That's remarkable," said Arthur.

"Just a personal quirk."

"No, not at all. It's a philosophy." He sat down and looked at me. "Had I thought of that I'd not have joined the parade. Now I feel like a trespasser in telling the tale." He paused to extract a cigarette from his case and light it.

"You were saying," I said.

"Well, nevertheless, there I was standing at the base of the fence watching the hurly-burly. There were three ballplayers near second base, engaging in some sort of contretemps, and of them, the Giant——"

"McGinnity."

"Whoever. He turned and threw the ball in my direction."

"You caught it?"

"Don't be absurd, I could never catch a baseball with any facility. A man near me caught it, and I paid him twenty dollars for it on the spot. I wanted it for you. It was the ball Mathewson pitched with, after all."

"And then Evers came running up to you and demanded the ball."

"Nothing of the sort. No ballplayer came anywhere near me. I put the ball in my pocket. I looked for you, first in the box and then near the clubhouse, but I couldn't find you. I did happen upon our clients, however. Did Eli just shrug them off to scramble after his winnings? I showed them the ball, you may ask them if you wish to corroborate my story. I've had it to this moment. This ball, Jack, the game ball. By the literal rules, as I understand them, the game is still in progress. Merkle may not have touched second, but neither did Evers, not in possession of the game ball."

A mad notion occurred: the rousing of O'Day, a rush to the Lambs Club, a midnight ride with Merkle to the Polo Grounds, a ceremonious touch of the base . . .

"Evers had a baseball, I saw it," I said.

"I don't doubt it. I imagine he has access to several dozen. However, he did not have this ball. Here, have it. It's yours."

I took it and held it as Mathewson had held it, two fingers across the seams. The stitching was specked with white; I remembered the puff of lime Merkle's hit had raised.

"Jack, hear me well so that you cannot misinterpret my words.

A man who will bet any significant sum on such a vagary ought not to be trusted. A man who will disturb a very fragile and vital relationship with a client for the sake of a wager ought not to hold a position of responsibility in our firm."

"What would Eli do?" I asked.

"His needs will be seen to. He's my brother as well as yours, he's family. Nor can I pretend that I am yet able to assume his duties. But between Uncle Sid and myself, we have to work something out."

"Have you talked to him?"

"To Eli? I'm not Billy Sunday."

"I meant, have you talked to Uncle Sid?"

"Uncle Sid and I talk between the lines. We understand one another. We recognize what's good for business."

"Doing in Eli is good for business?"

"Nobody's doing in Eli, save Eli himself. He's not acting responsibly. We all have our responsibilities, Jack."

"To poor Riva, may she rest."

"Indeed." Arthur rose, buttoned his jacket, and led me to the front door. "Please thank Edith for her hospitality. I had no idea I'd keep her so late. Good night."

"Good night, Arthur."

"By the way," he said as he opened the door, "do you know what it was that did poor Riva in?"

"She was sick."

"Yes. Food poisoning. A puff of pastry gone bad. She knew it had been lying there for weeks, but she couldn't help herself."

Rumor ran rampant the following day. Hour by hour the game was won or lost or tied once more; strange day, as if the contest were yet in progress, a suspense that overshadowed the undue fact that the Cubs and the Giants were to have at it again that afternoon at the Polo Grounds, though for what stakes none could say. Half the newspapers counted Wednesday's confusion a Giant victory and listed New York a game ahead in the standings; others accepted "tie game" as the immediate if not final result and repeated the previous day's listing, with Pittsburgh half a game closer to the top by dint of their victory in Brooklyn. There was a death watch at the Lambs Club, where John Brush was abed and not to be moved. At the league office Harry Pulliam brooded over Umpire O'Day's report and announced only that he had no announcement to make.

And at the shop there were the Christmas specialties, half a dozen finished designs required by the next morning. I left a note on Eli's door willing my game ticket to Ruby and locked myself into my office to work without interruption. By five o'clock three of the drawings were done, and I went to dine at a nearby tavern where I could follow the ballgame by ticker. When I entered Hooks Wiltse had a five-run lead in the seventh inning; by the time my sandwich was made, cut, and half-eaten the score was five to four, Wiltse was gone, and Mathewson was pitching in relief. He held the line, and the Giants won a certain and unalterable victory. The sports in the tavern joked about it as they traded money.

I saw no one in the office when I returned to work, but there was a light shining under Uncle Sid's door. An hour later I heard his steps approaching, and then the rap of his fist at my door.

"Gott sedanke," he said when I answered. "Somebody still works at this place. What you got doing?"

"The Christmas specialties, Uncle Sid."

"The specialties! What's it, August still?"

"They're due tomorrow."

"It's not too late? Reuven can still ship in time?"

"So he says."

"Wonderful. Oh, Yakov, the letter. Did you do the letter to Matson yet?"

"No, I haven't. I wasn't to write it until after the World's Series, remember?"

"Forget it. Don't do none at all."

"No? You changed your mind?"

"I'll tell you the story." He sat, removing his felt hat; a small yarmulke, as permanent a feature as his mottled skin or gray crown of hair, stayed atop his head. "I'm looking at this picture of Matson with the collars, Pearl the brand, and I say to myself, Pearl! Don't I know a guy Sam Weiskopf long time ago, cuts collars on Essex Street, wife named Pearl? No, I says, couldn't be. But then I think, maybe we get Matson in our picture, maybe Sam Weiskopf somewhere looks at the magazine and says hey! Don't I know Pincus once, a jeweler? Same thing! So I call on the phone Pearl Shirt Collars, I say maybe you got a Sam Weiskopf there. They say, got a Sam Weiskopf! He's president! So I say this is president Pincus, you give me president Weiskopf. He comes on phone, and I say hello president, he says hello president. Make a long story short, he's the same guy! Named the company after Pearl, that's nice. Maybe I should do for Riva, may she rest. So I get twenty minutes his life story, he gets twenty minutes mine, and then I say, hey, Sam, how you get Matson in your picture? You got a salesman knows him? He says no, he got a vice president for advertising. A vice president! And for advertising! Very big company.

"So I say, he knows Matson, right? Sam says, wrong. You don't got to know Matson. You got to write a man Frost, a lawyer, in Philadelphia. Frost handles everything, the whole deal. Sam says you go to Matson direct, you don't got a chance. Frost is the man. So I say, Sam, this Matson, he sells collars? And Sam says, a record year! A record year! So I say congratulations, and I hang up before maybe he asks me to dinner and already I'm married to his widow sister with the mole right next to the eye.

"So today I see your brother Caleb the lawyer. I tell him write to Frost, because to tell you true, Yakov, good as I am speaking English I don't write it so good. Caleb I make my vice president for advertising. He writes the letter, don't you. Forget about it. Just draw. That's nice?"

"That's fine, Uncle Sid. I'm relieved, really."

"I think so. You been worried a lot lately. It's Edith, maybe? She showing the pregnant yet?"

"Quite a bit, Uncle Sid. She's due in ten weeks."

"She's okay?"

"She's doing very well, thank you."

"Then what's the big worry I see on your face? Is it just this baseball?"

"That's part of it. Would you like some tea?"

"A *bissel.*" He held the cup in his fingers, disdaining the handle.

"Always in your head it's the baseball, Yakov. I remember you were going to run to Australia to play the baseball."

"Altoona, Uncle Sid."

"Same thing. Now, tell me true today, you got Edith, you got the baby coming and more later I hope, you got a nice home, you got a job for your talent. Are you glad you did the right thing, to work for a living and forget the baseball?"

"The right thing. If I did the right thing, I'm glad of it, yes."

"You did. How old are you, Yakov?"

"Twenty-seven."

"Twenty-seven, you're still young, twenty-seven. You got a whole life yet. In the baseball you'd be an old man, no?"

"Well, coming on middle age, anyway."

"See? You go into the business, it keeps you young."

"I'd never looked at it that way."

"The baseball's good, it gets you out in the fresh air, Eli says it helps sell the buyers. But it's not life, Yakov."

We sipped tea together. "What do you think of Arthur?" I asked.

He took the question seriously, answering in Yiddish. I listened carefully, for I was out of practice with the tongue.

"Avram. Well, I can't pretend to know him well. I've always been told he's brilliant, so I suppose I have to believe it. What's more, he's impressed me by coming to me man-to-man and asking for a job. That was hardly the case with you, or even Eli for that matter. Not that I regret it, but you didn't exactly beg me for a place, did you? No, it was your momma. So I gave Avram credit. And the fact is I need him, Yakov. The more I worry about Eli, the more I need Avram."

"You worry about Eli?"

"Don't you? I know you'd never tell tales on a brother, Yakov, but you should have told me what he's like on the road."

"It's always seemed harmless to me, Uncle Sid. I suppose I was used to it. Besides, it takes two to make a bet. The clients aren't children."

"Talk English, your accent is awful."

"I'm sorry. Arthur saw it all with a fresh eye, I suppose. He does seem to have an insight."

"He does. He's a bit overeager, but I can hold his reins for a while. And it's true, I need him. If I didn't have him I'd have to hire someone who isn't family, and I'd hate to do that. At least when family steals, the money stays in the family." He laughed and reverted to English. "So now I got three of you. Four! Caleb, my vice president for advertising. You want to be vice president, Yakov? Comes the baby, I make you vice president. And a thousand dollars more, too."

"That's very generous, Uncle Sid. Thank you very much."

"Welcome. Now, finish the specialties, go home, get a good night's sleep. More baseball tomorrow?"

"A double-header, Uncle Sid."

"That's nice, two heads are better than one. Good night."

The Giants faced four double-headers in a week. For the first of them they unveiled their "$11,000 Beauty," a spindly left-hander name Marquard, purchased for that unprecedented sum from the Indianapolis club of the Triple-A leagues. The Beauty withered in five innings, trounced in a lopsided loss. The Giants made it closer in the second game, but with no arms, no legs, no bats, they folded. Their lead over the Cubs was a single percentage point. On Saturday Mathewson took hold of the opener and won his thirty-third victory (thirty-four, if Wednesday's disputed result were added); in the second game, for the first time in a week, the club won without him. On Monday they rode with luck, coming from behind on two ninth-inning errors; on Tuesday Mathewson won the first game of yet another double-header, but for the second game the Phils sent out a youngster named Coveleski, as fresh to the league as Marquard but far more beautiful; he pitched a shut-out. On Wednesday I sat next to Eli and watched him wring his scorecard to tatters; we won, two to one. On Thursday Eli followed the club to Philadelphia; I spent the afternoon watching the tavern blackboard. Mathewson won the first game of one last double-header, but young Coveleski bested Wiltse in the second game. The Giants' percentage was .635, Chicago and Pittsburgh were tied at .633. My Friday visit to the tavern was brief, as the Giants scored seven runs in the first inning and coasted home. But Three-Fingered Brown pitched the Cubs to victory in Cincinnati, and Pittsburgh beat St. Louis twice to vault to a half-game lead.

On Saturday Mathewson pitched for the third time in five days, head to head against young Coveleski. He couldn't lose—but he did, by the narrowest margin, while the other contenders won. It was out of our hands now; Pittsburgh led the Cubs by half a game, the Giants by one and a half. The two leaders had a head-to-head match to play on Sunday while we faced three games against Boston, rained out earlier in the year but now necessary to play. Looming over all was the unresolved "Merkle" game, as everyone now called it. Pulliam's indecision had hovered over the race for ten days. Surely he hoped for a Pittsburgh victory on Sunday which would clinch the title and make the matter a grand irrelevancy.

Ten thousand enthusiasts gathered at the Polo Grounds on Sunday to watch an electric board blink out the play-by-play of the western contest; the event offered the ironic spectacle of a Giant crowd hollering itself hoarse for the Cubs. The Giant players sat in the press section, shielded from the crowd by a cordon of police. None were disappointed; down went Pittsburgh, and everything waited on the New York–Boston series. On Monday morning Pulliam finally published a declaration. The disputed game stood as a tie; O'Day was vindicated, Chicago's protest and petition for a forfeit denied. Some guessed that as dearly as Pulliam would have loved to steal the game from McGraw, he feared that the word would kill John Brush, still abed at the Lambs Club. The Giants were offered a choice, contingent upon beating Boston three times: they could have a best-of-five playoff with Chicago or a single game, winner-take-all. McGraw passed up three certain sell-outs at the gate and opted for the one game; he held Mathewson out of the Boston series to have him ready for the Cubs.

The Boston games drew meager crowds. Perhaps the city was exhausted by the dervish dance that had gone on too long by days; perhaps the outcome was preordained, for nothing else could crown the season but a final face-off against Chicago. Ames won on Monday, Wiltse on Tuesday, Ames again on Wednesday. Twice the Giants built huge leads, and out of the lineup came

Bresnahan, Donlin, and others of the list McGraw called "the identified dead." But Merkle stayed in, a gaunt scarecrow helpless at bat, as if he feared the basepaths.

With the final out of the final Boston game the city came alive to the race once more. Fans queued up for an overnight vigil at the general admission gates; the event became a raucous dusk-to-dawn festival, and writers worried that the goings-on would rob sleep from those Giants who now lived on the Heights above the park: Mathewson, Donlin, Bresnahan, McGraw himself. Certainly the neighborhood never slumbered that night, but neither did the Cubs, whose midtown hotel was surrounded and besieged by fanatics with horns and noisemakers. There was an ugliness emerging from the city's heart, born of three months of anxiety over the daily battles. I saw it first on the train to the Polo Grounds: two youths were scrawling obscenities about the Cubs on the walls of the car, and when the conductor confronted them they pitched paint in his face and escaped, laughing meanly. More astonishing was the laughter of the passengers and their cheers for the roughs. The Polo Grounds station looked like the aftermath of a New Year's revel; broken bottles, chucks of ribs and chicken bones, swirls of wind-swept newspaper, and a dozen foul drunken survivors coated with piss and puke. Thousands of furious turn-aways howled their disappointment and beat at the ballpark gates; a company of mounted police broke upon them, and there was blood on the stadium walls. It was nearly worth my life to get to the boxholders' entrance; twice my pockets were ripped by picks who disdained a light-fingered touch.

Inside the park the mood was equally angry. McGinnity and Chance came to blows over practice rights to the field; Eli thought it a stratagem to remove the Cubs' player-manager from the lineup, but if it was it was too obvious to succeed. Pulliam, who'd needed a heavy police escort to reach his field box, fined both players on the spot but allowed them to stay. As Chance turned away a beer bottle crashed against his neck, and he bled. Heavily bandaged, he came to home plate to exchange lineup cards with McGraw. At that moment there was a great shuddering noise from right field. The fence collapsed, and thousands stampeded onto the grounds. First the mob was threatened with fire hoses, but more than a threat was required; they had actually to be used, directly and brutally, and I saw men hurled about like fish in a furious surf. I was horrified, but many of the grandstand crowd laughed.

The worst had yet to happen. As Mathewson took the game ball in hand there was a terrible cry, and two men, locked together in arms, plummeted seventy feet from the grandstand roof to the bleachers below. The sight of them lying in a pitiful and ungainly heap was too much for even this crowd, hungry as it was for sensation, and in an awful quiet the poor dead men were wrapped in blankets and carried away by the police. In the delay other officers climbed to the roof and cleared it of hundreds of people. Never did an umpire's call of "Play ball!" sound so dreadful and empty.

The game bid against the terror of the afternoon. After Mathewson put the Cubs away in the first Tenney led off for the Giants; he needed a cane to climb the clubhouse steps, but it was beyond Merkle to play this day. Chicago had no pity: Pfeister hit Tenney in the leg with his first pitch. The veteran half crawled to first base. Herzog walked; we thought perhaps Pfeister was more unnerved than cruel, but then he struck out Bresnahan. Kling dropped the third strike, and Tenney, his face a riot of pain, dove safely into third. Donlin, the fourth Giant hitter, the third Giant cripple, slapped Pfeister's pitch into right field, and Tenney wobbled home. Pfeister was undone. He threw four balls to Seymour, arguing each call, and when Chance came to the mound to consult him the pitcher was clearly weeping. The Cub manager gently guided him toward the bench and signaled to the bullpen.

Across the yellowing grass strode Mordecai Peter Centennial Brown, he of the cut finger and the darting curve. It was just, it was just: a deformed, three-fingered devil of a pitcher would turn this diabolical day against the Giants and shut the rally down. It was just that Cy Seymour, searching for the ball against an unfamiliar October sky, would turn Tinker's third inning fly into a triple, and that Joe would score on a bloop hit. The same justice required that Wildfire Schulte stand in with two on and two out. Two weeks before, Mathewson had struck him out three times. But for Merkle, he'd not be batting now; but for Merkle, we'd be entrained to Detroit for the World's Series, and two anonymous men would be alive in the excited city, arguing their club's chances to the music of a ticker. Instead they were crushed and dead. Of course Schulte doubled home a run to put the Cubs ahead; of course Frank Chance followed with a drive that came to earth exactly where Merkle's hit had landed, scoring two more runs. The fans, witnessing another death, were silent; the Cubs led, four to one.

That was the score in the seventh, when Devlin finally touched Brown for a hit. McCormick moved him along with a single. Bridwell—Bridwell's hit had won the Merkle game!—Bridwell walked. The bases were loaded and none were out, with Mathewson to bat.

Mathewson was to bat—but McGraw stepped in his way. Instead, Larry Doyle limped out of the dugout swinging his stick, and Mathewson turned and disappeared from sight.

I stood up, buttoned my topcoat against the autumn chill, and stepped into the aisle.

"You're not leaving, sport! Bases loaded! Tying runs on!"

The crowd groaned, and I turned to see catcher Kling settle under Doyle's pop foul. Slowly, careful of the steep descent and holding the rail firmly, I walked down the ramp. I heard no echo of the song that Eli and I had sung two weeks before. From the base of the ramp I glimpsed Tenney swinging at Three-Fingered Brown's pitch, heard the connection, watched Devlin tag at third and dash down the line with a Giant run. I continued along behind the standees, five deep, heard their unhappy cries at a third Giant out, and caught sight of the teams changing sides. I was cold. I stopped to buy a cup of hot chocolate from a vendor; it burned my lips. I took a glance at the field. Mathewson, sweatered, was walking slowly across the outfield to the clubhouse. I watched him climb the steps, and then I left the park.

October 16, 1912—at Boston

											R	H	E
NEW YORK	0	0	1	0	0	0	0	0	0	1	2	9	2
BOSTON	0	0	0	0	0	0	1	0	0	2	3	8	5

BATTERIES: New York, Mathewson & Meyers;
Boston, Bedient, Wood (8) and Cady.
Winning pitcher: Wood. *Losing pitcher*: Mathewson.

THE WORLD'S CHAMPIONSHIP SERIES

	W	L	PCT.
Boston (A.L.)	4	3	.571
New York (N.L.)	3	4	.429

Boston wins best-of-seven series, four games to three.

FOUR

THE POLO GROUNDS fire that burned for a day and a night in the spring of 1911 began with an overturned brazier at a vendor's stand, but the effect was so out of proportion to the cause that rumor assigned the spark to Bolsheviks or Chicagoans. McGraw suspected Ban Johnson. John Brush, shrunken and frail, watched from his wheelchair and wept as the brigades vainly battled the flames. It ended with a charred ruined head in the hollow below the Bluff, and a competition among editorialists to give symbolic meaning to the event.

It was never doubted that the park would be rebuilt; plans were drawn while the embers still smouldered. In the interim a place had to be found for the Giants' summer of games. The league schedule made it impossible to share Washington Park in Brooklyn, and a fantastical suggestion to erect great towers of arc lights there to permit play at night was dismissed out of hand. My father-in-law carried to Brush an associate's generous offer of his private polo field on Sands Point, but the invitation was withdrawn when the philanthropist came to understand that the public must be admitted to watch the games. At last the awful alternative was grasped, and a humble Brush begged the Highlanders of the American League to permit the Giants use of Highland Park. It was a delicious moment for Andrew Freedman and his partners, and their terms were extortionate, but there was no choice in the matter. Freedman and Brush signed the compact in a grand City Hall ceremony while McGraw and the club were on the road.

It was strange to watch the Giants on the foreign turf—I attended only one game—but the youngsters flourished there, especially Merkle, who frolicked like a pardoned convict and led the club in run production. Snodgrass took over center field and yet a newer man, Josh Devore, claimed left; Larry Doyle showed surprising power and with shortstop Fletcher provided a solid middle defense; Herzog played wherever his glove could do the least damage. McGraw released Bresnahan to manage the St. Louis club and gave his place to a muscular copper-skinned Shoshone from the Indian missions of California; his name was Meyers, and of course they called him Chief. The wounded Donlin retired, but lest the city's bartenders mourn too greatly McGraw purchased Bugs Raymond who, when he could stand, could pitch a pretty piece. Marquard convinced many that his $11,000 price tag was a bargain. McGinnity was gone, and so was Dummy Taylor, and for young Crandall McGraw invented a role: If the team trailed in late innings Crandall would pinch-hit for the pitcher, for he had a strong bat, and then take the mound in relief. Hooks Wiltse and Red Ames faded. Still Mathewson reigned.

New as the uptown park was to me, it was familiar ground to Eli. With his fortunes at low ebb after the Merkle game he'd found a new and resuscitating world there. He courted and cultivated Hal Chase, who was unarguably the best player on the roster of the Highlanders (or Yankees, as the newspapers came to call them, for the shorter name better suited their headlines). Chase had played a winter in the outlaw leagues west of the Rockies, then came back east with a promise from Andrew Freedman that ought never have been sworn: if Stallings, the latest manager, didn't win, Chase could have the reins. Chase, famous for his fielding, proceeded to lead the league in errors two years running; Stallings didn't win. When Chase succeeded him in 1910 Eli bought the player an Olds roadster to celebrate his elevation. "Those Yanks are a sure thing under Prince Hal, win or lose," Eli would say with a wink. "Only a fool would pass on a sure thing."

Eli pressed us to buy an endorsement from Chase on the Mathewson model, but Arthur surveyed opinion among the buyers and advised against it. There were mean rumors afloat concerning Chase, and Eli's growing bankroll seemed to confirm them. The matter was tabled, and I was relieved. It had been difficult enough to design the Mathewson line. Although I would have carved

Mathewson's statue for the spire of a cathedral, I'd found it impossible to express myself in studs and tie pins for his endorsement. My initial drawings satisfied no one, least of all myself; Arthur thought them too traditional, Eli too unusual, and Uncle Sid too expensive by half. Finally I took the seams of a baseball as my motif and worked the pattern into a set of articles that Harry Stevens' concessionaires might have sold at the Polo Grounds for nickels. In these I took no pride, and I wondered if Mathewson's lawyer, the man Frost, felt as corrupt in arranging the endorsement as I did in designing the line. We'd dealt with Frost exclusively; I couldn't know if the pitcher was even aware of the agreement, and I hoped not. As the project grew in scope so did my low opinion of it. Hundreds of stores signed up to carry the line, showing it in bright displays of Arthur's devising which featured a life-sized cardboard cut-out of the hero. To decorate Mathewson entirely, as if he were a knight errant, and then to offer his shield and blazon for sale so that every man might be a Mathewson! It was too ridiculous.

The line was introduced under a new corporate insignia. Pincus Jewelers was no more; Collegiate Jewelers was born. This was Arthur's doing, as was the advertising that appeared in national magazines and local newspapers quoting Mathewson thus:

> "I've been wearing Collegiate Jewelry since my own college years! Collegiate covers the field like a ballhawk, from the most formal fashions to the snappiest designs for the real sport. You'll never strike out with Collegiate Jewelry, for yourself or your sweetheart!"

To every complaint I made of the text Arthur had a ready answer. I called the first phrase an outright lie; Arthur reminded me that Mathewson had been enrolled at Bucknell when he'd received the ruby ring. I insisted that it had been a gift; Arthur noted that he'd written "wearing," not "buying." The expression "snappiest" was unknown to me, but Arthur insisted it was apt. Finally I argued that it was Pincus jewelry, not Collegiate, that Mathewson had been wearing. "A technicality," said Arthur. "I hardly intend to use our advertising to trace the etymology of the corporate name." He carried every point.

He had yet greater plans in mind, and memoranda flowed from his desk. Citing statistics from a dozen cities, he complained that only a small percentage of shoppers found their way to jewelry departments on upper floors; he advised that we push our client stores to relocate the counters on the main shopping floor, hard by the most trafficked entry, and in every case he knew where this was. Against their refusal he examined sites for stores of our own, envisioning them as exclusive carriers of our line. He urged us to expand into the manufacture of watches. He argued that we ought to open factories where labor was cheaper than New York's extravagant standard of fourteen cents an hour, and reduce costs by shipping merchandise in parts to be assembled locally. He begged for the New England territory as a proving ground for these innovations. Uncle Sid took the suggestions under advisement, allowing some, vetoing others, while Eli kept his own counsel. He might have been blind to Arthur's ambition; he might have read it and welcomed it. There was no telling.

I was excused from the annual western swing. Eli made the suggestion, saying that a husband and father should stay at home, and Arthur echoed it with the flattery that the family's artist ought not have to bear the drudgery of the trip. I didn't argue. Already I'd separated myself from the daily workings of the shop and Arthur's incessant campaigning; I worked at home now, in an office designed for efficiency's sake. It had but three items of decoration: on the wall were my drawings for the ruby ring of '01 and the triple diamond of '05, and on a shelf sat the Merkle game ball. My son discovered it and constructed an ingenious ladder to reach it; it became a favorite toy. I judged he was ready for a ballgame, but I wouldn't take him to the Giants' temporary home; for my son's initiation only the Polo Grounds would do.

It reopened with appropriate ceremony in August of 1911. Officially the park was redubbed John Brush Stadium, but never was a name so thoroughly ignored; they might as well have tried to rechristen Sixth Avenue. There was seating for forty thousand, and room for many more to stand. The bleachers surrounded the entire outfield, and the horseshoe grandstand was extended another seventy yards on either side, two decks high. Every single grandstand seat had a back and arm rests. The luxury was astounding.

Better yet, the club was worthy of the resurrected park. Giant youth was finally an advantage, for Chicago and Pittsburgh were slow with age. Evers played fewer than fifty games that summer, and Wagner had his lowest batting totals in a decade. The three old rivals were closely bunched, but the Giants were on the upswing. McGraw called the emerging club better than the '05 champions, better even than his beloved Orioles of the 'Nineties, and he ran them with furious abandon. They beat Pittsburgh on the first of September to vault into first place and never relinquished the lead, clinching their first pennant since '05 with a week left to play.

In that last week, with the crowds less numerous and passionate, I took Matthias to watch Mathewson tune up for the World's Series. At the turnstiles a great, ancient bear of a man checked my pass. He was indifferent to my son's excitement, and when Matthias playfully reached for a dirty handkerchief that drooped from the man's uniform pocket he slapped the child's hand away. In contrast a program vendor steps beyond chucked Matthias under the chin. "What's the matter, Amos?" he called to the ticket taker. "Got somethin' against kids? I weren't much older, first time I seen you pitch."

I turned to see the old man shrug. Rusie! The idol of my youth, grown fat and gray. His eyes were dull, and his bulk stretched the fabric of his jacket. I walked back to him.

"You pitched the first game I ever saw here," I said. "My brother and I watched from the Bluff."

He took my offered hand, but his face remained a blank. "How old's the boy?" he asked.

"Nearly three."

"He'll need his own ticket, time he's six." He turned away to read the indicator on the turnstile.

I took Matthias in my arms and carried him up the ramp. As we stepped into the grandstand aisle I saw the blond pitcher trotting down the clubhouse steps, and people came out of their seats crying "Matty! Matty!" Matthias bubbled over. "They know me!" he thrilled. "They know me!"

Matthias was delighted to find his Uncle Eli in the box; I was something less enchanted to discover his guest, Hal Chase. Matthias, though, immediately took to the Yankee, whose huge hands held him gently.

"Did you see Amos Rusie at the gate?" I asked Eli.

"Sure, he's been here since the park opened," Eli said. "Matty's responsible. He read an item about Rusie's farm going bust and showed it to Mac. They picked him up in Cincinnati and tried him as a pitcher's coach, but what could Rusie teach? Just rear back and fire, that was his style."

"It's sad to see him down there," I said.

"It's a damn sight sadder to picture him on his front porch, watching the weeds grow," said Chase. "He ought to be a lesson to those boys on the club. Grab it while you can, 'cause no one else is watching out for you. Take Matty. He's been putting it away for ten years now. Smart fellow. He'll pick up your check and make you think he's the greatest guy in the world, and then he'll win it back double at checkers."

Matthias was indifferent to the action on the field, but other aspects of the Polo Grounds enthralled him: the variant crowd, the changing numbers on the scoreboard, the pigeons swooping over the roof and roosting in the wrought-iron girders of the grandstand. Finally he fell asleep in Chase's arms, which charmed the player. I offered to take him, but Chase insisted he was no bother.

"Do you like the Giants in the Series?" I asked.

"Nope. Philadelphia should beat their brains out. You saw what they did to the Cubs last year. They'd have taken it in four games, but Connie wanted another home date so they blew the fourth game in Chicago and ended it in Philly. Connie Mack's no chump, he can count."

"And your club? Will it be in the race next year?"

"Shit if I care. I've been informed that my services as manager are no longer required. Hell, I lasted a year and some with Freedman, better than the average. I put away a little. I'll be damned if I end up like Rusie."

"Where will you go?"

"I'll stay right here. Don't much matter who's running the club, and there's more money floating in New York than in any three other cities, 'cept Chicago. I'd be a fool to leave." Matthias stirred, and he soothed the child by stroking his hair. "Good-looking kid. I probably have a few of my own out west. Hey, Kappy, that reminds me. What about that ring?"

Eli turned to me. "I wonder, sport, if you could do something in leather with a half-inch band, and stud it with baguettes. Make it about an inch and a half in diameter——"

"Thicker than that, Kappy. What do you take me for? You think I'm your size?"

"Measure it yourself, Hal, and don't brag about it. You don't want it slipping off when you're in high gear."

"I'll see if the lady has a tape measure. She's got just about everything else in that drawer, 'cept a ring like that. I want to be her diamond-studded stud."

I muttered that I'd never worked with leather, which set both men laughing. "God damn it, Kappy, you're all promise and no delivery. Just like the lady says."

"Jackie will be too busy in any case, Hal. Rings for the Giants, rings for the Athletics—we've got a contract with the National Commission. We won't be through the list until Christmas."

"You do spread it around, Kappy. Say, how do you come up with a new idea every year? I've seen those rings, I wish I had one. They're damn good."

"There's a basic model, the changes from year to year are microscopic," I said. "It's a matter of economy. We try to use the dies we've already cut."

"King Arthur's idea," said Eli.

"He's right," I said, "and not just from the money side. With all the work we have I couldn't possibly come up with brand new designs every year. The Mathewson line is the only thing I really put my mind to."

"Does he sell?" said Chase.

"Does he sell! You should make so much, Hal. Over the counter, I mean."

"Don't matter how it comes in, Kappy. It's all the same going out."

"You're a philosopher, Hal."

"No. I am a professional ballplayer. 'Professional' means only one thing. You do it for money. But the bastards don't let you go where the money's best. They strap you with the reserve clause and tie you up for life. The National Commission may be bread and butter to you, but it's cost me thousands. If I'd signed two years before I did, when the leagues were at each other's throats, I'd have made three times the money. But shit, by the time I came up salaries had been cut to nothing. I hit twenty points over three hundred my first full year, and that winter they sent me a contract that knocked me down five hundred dollars. 'Sure you did all right,' they said, 'but the club didn't win.' As if it was my fault! I'd been busting my ass all year. Well, that's when I saw my way clear. I'm a professional ballplayer. I do it for money, and if there's more money in losing than in winning, shit if I care."

Chase and Eli left early, chuckling over a pair of quail they were to meet at the Ansonia. I wanted my son's visit to be complete, and at game's end I led him along the grandstand aisle toward the clubhouse, but the old path was barred. It would have been easy to walk across the outfield to the players' stairway there, but that was not my practice. With Matthias on my shoulders I explored the labyrinthine passageways beneath the stadium and finally emerged on the old veranda. At the clubhouse door the cop, whose name was a distant memory, stopped me and studied my face.

"Hold on. You're Kappy's brother, ain't you? The one who does the rings?"

"Yes, that's me."

"Wait here." He closed the iron door, leaving my son and me alone on the veranda.

"They know you too, Daddy," said Matthias.

"I guess they do."

The door opened and the cop stepped out. "Mac wants to know if you got them rings," he said.

"What rings?"

"Seven rings he says he ordered from you. He says you don't get in 'til he sees them rings."

"My God, that was three years ago!"

"Ain't none of my business when it was. You don't get in 'til Mac sees them rings. Where are you going, sonny?" Matthias was peeking in the doorway. The cop put his massive hand on my son's shoulder.

"Don't touch him!" I said.

"Keep him from where he don't belong," said the cop, pushing the boy toward me. His small face was pale, his eyes wide with fright. I picked him up and hugged him, and now, safe in my arms, he could cry.

I was generous with my World's Series tickets; anyone who asked could have them. I didn't see Mathewson's first-game victory, nor did I follow the club to Philadelphia to see Marquard beaten and the Series tied. On Tuesday morning we had a visitor at the shop: the lawyer Frost, come to tour the facilities that were enriching his client. He insisted that Arthur and I be his guests at the game that afternoon; we sat in the field boxes close by the Giant dugout. Frost was a small, square man of the older generation; he'd handled the Mathewson family's legal affairs for years, and was the first person I ever heard call the pitcher "Christopher." Frost regarded Mathewson's name as an income-producing property, and the only limit upon its use was his own sense of propriety. He told us that he'd turned down an endorsement of a patent medicine when the applicants proposed the slogan, "Guaranteed to make your every ailment 'fade away'!" While I watched Mathewson duel Jack Coombs under a darkening sky Frost received a variety of callers; all the talk was of figures and percentages. It was a veritable auction.

The year before, when he'd beaten the Cubs three times, Coombs had been dubbed "the American League Matty"; he too was a collegian and a control artist. Mathewson built a Giant run in the third with a base hit, and protected it to the ninth; it was one to nothing when Philadelphia's oddly-proportioned third baseman, Frank Baker, came to bat. Above the waist he was whippet-thin, with elongated arms that gave him huge range in the field; below, he was stocky and powerful, strong enough to have led his league in home runs with nine. He'd hit another the day before to beat Marquard. Now he got under Mathewson's fastball and sent it high down the right field line. The Polo Grounds oval that created a generous center field had very near walls down either line, and the ball made the seats by a fraction. The game was tied. Mathewson seemed unfazed, but his teammates were undone. They couldn't touch Coombs in the bottom of the ninth, and when they batted in the tenth both Merkle and Snodgrass were thrown out on the bases. In the Philadelphia eleventh Baker hit a grounder to the egregious Herzog, who stabbed it and promptly threw into right field. With runners on second and third Mathewson forced a bouncing ball to shortstop; it ran up Fletcher's arms, and the tie was broken. On the next pitch the baserunner broke for second, and both Doyle and Fletcher moved to cover the base; a ground ball rolled over the vacant infield to bring Baker home. The Giants scraped a run together in the bottom of the eleventh, but the rally fell short and the game was over. It left me exhausted, but the moneychangers at Frost's box were oblivious to it all. On the journey home Arthur outlined the dimensions of a new deal he'd struck with the lawyer, and when we parted he asked, as an afterthought, who'd won the game.

Then came the wintry weather. For six days a freezing rain prevented the fourth game. Too much rest was worse than too little; although Mathewson struck out the side in the first inning, he struggled thereafter and allowed three runs in the fourth inning; it led to his second defeat. The next day's game began awfully; Marquard gave up a three-run homer and the Giants were in desperate shape. But they rallied in the ninth to tie the game, and won it in the tenth on Merkle's scoring fly ball. That night at the Ansonia the pariah of '08 was toasted as a hero of Spartican dimension, but Hal Chase was there to repeat that Connie Mack could count.

I followed the sixth game on the huge board in Times Square. For the temperature as well as the mob it might have been New Year's Eve, but the party was a dub. Ames was chased in the fourth inning, and in the seventh the Giants collapsed in the field as the Athletics paraded around the bases. The Square was dawn-empty when the game and the Series ended.

In the spring a purported autobiography was published under Mathewson's name, entitled *Pitching in a Pinch*. That a ghost-writer had fashioned it was an open secret; Frost had sold the rights to a journeyman sportswriter named Wheeler, who gave interviews modestly acknowledging his role and swearing that no writer since the authors of the Gospels had borne a greater responsibility. If Christ had been as gentle with the Pharisees he'd have died of old age; not a line in the book could possibly offend. There wasn't a word of the broken contract with Mack or the hostilities of McGraw's early years, and with all the lessons on "inside baseball" there was nothing about the kind of pitching that had flipped Padden on his back in St. Louis and kept batters wary for a decade after. Only in the section on pitching under pressure did I find a trace of Mathewson's own voice, an echo of our supper at Mr. Sonnheim's club:

> It is in the pinch that the pitcher shows whether or not he is a Big Leaguer. It is the acid test. That is the reason so many men, who shine in the minor leagues, fail to make good in the majors. They cannot stand the fire.

The book had three printings, and a special Boy Scout edition sold out. Arthur was upset that there was no mention of Collegiate Jewelers, and Eli looked for his own name in vain.

I read the book in my workroom, and when I was done I stared at the drawings on the wall, envying their youthful energy and clean direction. In creating them I'd felt myself an instrument of Mathewson's own genius, but like Eli I'd staked my fortune on the vagaries of the game, and through the years of accident and frustration I'd become like Wheeler, exploiting rather than glorifying the hero. I scanned my current work, and I was ashamed; I realized that I needed new inspiration and could not look to Mathewson to provide it.

I stayed away from the Polo Grounds in 1912. I apportioned much of my work among the growing team of apprentices and spent my days in the museums and galleries. I asked Edith to plan a European tour for the fall. I sought out the attractions of the city that I'd always ignored in favor of a ballgame, and discovered that Washington Square was a public green of bright and busy delight, and that Little Italy to the south was a gastronomic adventure that feasted a new saint every Saturday. I took Edith to Coney Island's beergardens and saw singing waiters in black-face imitate Jolson and Wheeler, and then I found the genuine articles at Hammerstein's and the Winter Garden. I heard the young Galli-Curci sing and saw Strauss conduct, and I clapped to the beat of Sousa's baton at the Central Park bandshell. I watched Matthias play.

In August, entirely by happenstance, he saw me pitch. We spent the month with my in-laws on Lake George, and on a day trip to Warrensburg we laid a picnic on the lawn of the Congregational Church at the edge of the village green. Edith's mood was delightful. She joked that she'd never before dined *al fresco* and *ex cathedra* at a single stroke. When the locals began to choose up sides for a lunch-hour game she volunteered my services and advertised my rusted pitching skills. In a trice I found myself on a ragged patch of dirt that was more a depression than a mound, facing a youth of no more than fourteen who swung a bottle bat with exuberant energy. I walked him and two others in the first inning, but in the second I found my control by throwing harder; I struck out the last man. In the third I tried a curve ball, and it was a lovely thing. The amazed hitter backed out of the box and asked for a consensus ruling to bar me from throwing another. The matter went to a vote, which I carried by pledging to try no more than three in any single inning. At bat I was helpless.

In the sixth inning my pitches began to stand still over the plate, and I worsened matters by bobbling a bunt and throwing wildly to first. With two on base, the youngster with the bottle bat dug in against me. I threw with all my might and hit him squarely in the back. Everyone gathered around the boy, who recovered his breath, forgave me with a joke, and regained his feet as the church bell chimed the hour. For all the messiness of the last inning I heard the ringing as a *gloria*. When Edith made a teasing report of the game to her parents that evening I joined in the laughter, but the pain in my arm the next morning made me groan aloud. This earned sympathy from Edith, but none at all from her father, who for the rest of the summer delighted to call me "Lefty."

When I called at the shop in early September Arthur gave me assignments beyond the fixtures of the line. "We ought to do a commemorative for Marquard," he said, "and of course you know that young Tesreau pitched a no-hit game last Friday. You should do something for him as well."

"Are these Eli's orders?" I asked.

Arthur looked amused. "No, Eli's taking rather few orders these days. Now Marquard's the coming man, Jack, and I want to approach him for an endorsement. Nineteen victories in a row! Not even Mathewson ever did that. I'd like to make a down-payment on Tesreau as well. We have to look to the future. The championship rings are done, aren't they? If the Red Sox win I want them in the window at the opening of our Boston store."

"Ruby had them under way when I left. I imagine all that's needed is the engraving of the winning club's insignia."

"Of course. Remember, Jack, we have to make a strong impression on Marquard and Tesreau. What's wrong with your arm?"

"Is it that obvious? I pitched in a pick-up game in Lake George. I had a grand time, but I've been feeling it ever since."

"Did you do well?"

"Nothing worthy of a ring. Arthur, I'd like to give these Marquard and Tesreau pieces to young Mark. He's by far the most talented of the apprentices, and I want to see what he can do with the assignment."

"No, do them yourself." He turned his back and opened a book of accounts.

"That's rather abrupt," I said.

"As long as you ask, that's my answer."

"I've always had responsibility for assigning work."

"Then why ask me?"

"Because you told me directly to do these myself."

"Are we to have a jurisdictional dispute?" He turned to face me. "I want you to do them because they're important to the future of the firm. I thought I'd made that clear. The old order is passing, Jack. If Marquard isn't the league's best pitcher he soon will be. Moreover, he's touring the vaudeville circuit this winter and I want him wearing one of our rings. He'll be asked about it, and the publicity will suit us well. If you think that Mark can do a better design than you——"

"That's not the question. The point is that I haven't the impulse to design for Marquard and Tesreau. The Mathewson rings, those early ones, were never assigned or commissioned. They were done on impulse."

Arthur looked at me, snapped open his silver case and lit a cigarette. "You were made for one another, Mathewson and you," he said, shaking his head. "You think he's a kind of god, and I suspect that he shares your belief. Happily for him, he has the good lawyer Frost to deal with all the mundane and tawdry things in life, like endorsements and fees and checks such as we issued to him last year in the amount, may I say, of four thousand, seven hundred and sixty-four dollars and thirty-nine cents. You, poor man, have to lower yourself to design pieces which are not outpourings of pure worship. Well, I'm terribly sorry, but this is not an abbey and you are not a monk illuminating pages for the greater glory of God. This is a business concern which at present, for better or worse, depends on the endorsement of celebrities for a goodly proportion of its income, and if it is to expand and thrive—and I intend to see that it does—it cannot wait on impulse, and it cannot stake all on last year's hero."

I got up. "Then perhaps you ought not to depend on last year's designer," I said, and crossed to the door.

"Jack." I stopped. "Jack, I'm sorry. There's just so much pressure. So much depends on the Boston store. I understand your attitude, truly I do, and in other circumstances it would be laudable, but you must understand how unprofessional it is. You have responsibilities to the firm, and you ought to approach them as a professional."

"I remember hearing Hal Chase lecture on the subject of professionalism," I said.

"Chase? Eli's playmate? You won't believe what they asked your friend Mark to make for them."

"Touching Chase I'd believe anything. Arthur, I'll draw something for Marquard and Tesreau and I'll check with Ruby on the status of the rings for the Boston store, if you give me the date."

"October sixteenth."

"World's Series week."

"What a ball fan you are! It's also your wedding anniversary."

"I know that, Arthur, but I imagine it's the Series that concerns you."

"True enough. It's the day after the seventh game, if the Series goes that far. I couldn't hope for better than to have Boston and New York in the World's Series, and our store to open that very week. Have you seen the photographs?" He took an envelope from his desktop and handed it to me. I saw a riot of construction.

"It's very impressive. I'll be sorry to miss the opening."

"Miss it! What do you mean?"

"We're sailing for Europe on the sixteenth. You see, I really do remember my anniversary."

"Change your plans."

"That's impossible."

"Change them! You're receiving an award from the National Commission that day, it's part of the ceremonies."

"Arthur, we've planned this trip for months."

"Embark from Boston. You can attend the ceremonies in the morning and sail that afternoon, our local people will see to the arrangements."

"Arthur——"

"Jack, you have to be there. Mathewson's presenting the award."

I gaped. "He'll be there? Mathewson will be at the store?"

"Why not?" said Arthur, crushing his cigarette. "Christ walked among publicans and sinners."

While the Giants ran away from the National League pack the other race went down to the final weekend. There were three contending clubs: Philadelphia, the defending champions; Washington, a surprising challenger with little more to boast than its locomotive of a righthander, Walter Johnson; and Boston, the club McGraw had not deigned to play in 1904. I knew little of the American League and asked Eli's assessment. "Boston," he said without hesitation. "If Connie Mack's infield is worth a hundred grand, what would you give for Jake Stahl's outfield? A quarter million?" Hooper, Speaker, and Lewis: Boston hadn't seen such artillery since Bunker Hill. Harry Hooper and Duffy Lewis on the flanks were greyhounds at the chase, and Tris Speaker, no taller, carried twenty more pounds in muscle across his chest and arms. At bat he had no peer for percentage and power; nearly half his hits went for extra bases. In the field he played with his shadow over second base, contemptuous that anyone might hit the ball beyond his extraordinary range. There were others—Gardner and Stahl at the corners of the infield, .300 hitters both; Steve Yerkes and Harry Wagner (to ball fans, the "other" Wagner) at second and short, and strong-armed Bill Carrigan behind the plate—but the arm of arms belonged to Smokey Joe Wood. Mathewson at his finest never won so many games while losing so few as Wood did that year, and McGinnity in his prime was never so powerful. "Can I throw harder than Wood?" Walter Johnson asked. "Listen, my friend, there's no man alive can throw harder than Smokey Joe Wood."

They played before a fervent and uproarious pack that had organized itself into a grand fraternity, the Royal Rooters of Boston. These high-rolling Irishmen ached to play the Giants; they still felt the sting of McGraw's ancient insult, and when the Red Sox clinched the flag some two thousand of them carried their pennant celebration to New York aboard a chartered train. They tumbled into Grand Central on the eve of the World's Series bedecked with red sashes and top hats speckled with golden glitter, trooped up Broadway to the step of their own brass band and stormed the Bretton Hall Hotel at Eighty-Sixth Street to establish headquarters. Two hours later they paraded back to Times Square by torchlight, fortified against the evening chill by their passion and their flasks. They serenaded the amazed district and demanded a solo rendition of "Sweet Adeline" from their leader, the Mayor of Boston. His name was Fitzgerald—"Honey Fitz" to his devoted following—and his uncertain tenor was painful to hear, but his delightful face was a classical mask of comedy and his ebullience forgave his outrageous manner. The rally done, the Royal Rooters dispersed with currency of all denominations waving from their pockets and hatbands. They meant to cover every bet the city would offer. Bookmakers were charging seven dollars for the chance to make five on either club; New Yorkers demanded longer odds when McGraw announced that the rookie Tesreau would face Smokey Joe in the opening game. It was another of the little round man's moves that defied expectation but yielded to easy logic. If two thousand Royal Rooters could create an unnerving commotion in New York, what would thirty thousand do in Boston's ballpark? McGraw would hold Mathewson to face that mob and gave his younger pitchers the advantage of a home crowd. Marquard had been a midseason miracle, but Tesreau held the hotter hand down the stretch.

Fifty thousand jammed the Polo Grounds while hour by hour an apprentice brought me editions of the afternoon papers. For five innings the choice of Tesreau seemed inspired. He allowed not a hit, and the Giants gave him a two-run advantage in the third. Boston rallied to halve the lead in the sixth and overcome it in the seventh; the Giants trailed by two. In the ninth Merkle singled, then Herzog, then Meyers, but in the tumultuous moment Joe Wood showed enormous courage. He struck out Fletcher on three pitches, and on a full count he struck out Crandall. It choked the screaming mob, and it won the game for Boston.

Rail tickets to Boston were as precious as passes to the game. Less organized but no less cocksure than the Royal Rooters, the New York gang celebrated the certainty of Mathewson's coming victory as they sped through the New England night. Arthur and I had a different concern. We alternated guard on two huge crates of merchandise for the Boston store, exchanging a small black revolver at each relief. We supervised the unloading at dawn, and with an escort of private detectives we wheeled a rented truck through the still sleeping streets. The store was situated directly opposite the great Jordan Marsh emporium on Washington Street. By nine o'clock a crowd had formed in front of the store, and they cheered when we unveiled the championship rings in the show window. I saw people on the upper floors of Jordan Marsh frowning and shaking their heads; one raised a fist, and one of our workmen responded with an obscene gesture. Arthur took the man aside, cautioned him mildly, and slipped him a dollar bill. At noon we affixed a large banner across the storefront: GRAND OPENING/ GALA CELEBRATION/ OCTOBER 16. Under this was an italicized line: *Collegiate Jewelers/ We Adorn The Champions.*

When I heard the band approaching I automatically assumed that Arthur had arranged the show, but then I saw the sashes and top hats of the Royal Rooters and realized that it was a rump brigade on its way to Kenmore Square. Their extravagant buoyancy was seductive, and I enjoyed their hallelujah to the home roster:

Carrigan, Carrigan,
Speaker, Lewis, Wood, and Stahl,
Bradley, Engle, Pape, and Hall,
Wagner, Gardener, Hooper too—
Hit them! Hit them! Do, boys, do!

After lunch a young sales clerk took me to the Cunard offices in Boston harbor to arrange passage on the sixteenth. Service was impossible, for the commercial wire carried a pitch-by-pitch account of the game and took all the attention. The clerk had a shamrock encased in glass on his keychain, and he rubbed it at every close moment, muttering Gaelic spells. The thing worked terrors. Fletcher made an error in the first inning, two hits ricocheted off Mathewson's glove, and three Boston runs crossed home. The Giants narrowed the gap, but in the fifth Fletcher dropped a throw and Boston scored again. Once more the Giants rallied, and in the eighth they took the lead. But the shamrock had a sign on Fletcher: a two-out grounder went through his legs, his third error, and the score was tied. My business was long done —I had passage on a steamer departing at midnight on the sixteenth, and the clerk, who knew my sympathies, said, "Midnight! That's a sign!"—but there was no leaving the office with the game undecided and in extra innings. In the tenth McGraw used up his bench with pinch-hitters and pinch-runners and gained a run; Mathewson needed three more outs. The clerk put the shamrock on the floor and danced around it. With one out Speaker hit a long drive that might have been caught, but the new center fielder waved at it; Tris might have been nailed at third, but the new shortstop muffed the relay; he might have been out at home, but the new catcher fumbled the throw. It was a tie game, and I wanted to grind the shamrock under my heel. Another scoreless inning passed. The harbor hid in darkness, and a lighthouse beacon flashed on the horizon. The wire fell silent for a long moment before it clocked out the results: GAME CALLED DARKNESS FINAL SCORE SIX SIX.

Arthur was alone in the cluttered store. I saw the black revolver in his belt when he opened the door for me.

"Where are the guards?" I asked.

"A constable came by an hour ago and deputized them," he said. "They're expecting trouble. With the tie game no one knows whether to board the trains to New York, and with the crowds at the Southside Station . . . Have you eaten?"

"I'm not hungry."

"I am. Take the gun, I'll be back in an hour." He paused at the door. "They must add a game now, mustn't they? Perhaps they'll play on Sunday."

"Not in this town," I said.

"I suppose not. Then everything will be pushed back by a day, won't it? That means that if the Series goes the limit the store will open on the morning of the seventh game. We'll have all the players, invite the whole press corps. The publicity will be enormous. Can you picture it, both clubs slavering over the World's Series rings?" Laughing, Arthur stepped out into the sea mist that shrouded the empty street and closed the door behind him.

I put the revolver in my pocket and took my sketchpad from a counter. I switched off the lamps in the front room but left the window alight, to the advantage of the display there. I made my way to the storeroom in back and cleared off a workspace on a crate. I began to draw freely: tall-masted ships under full sail on stormy whitecapped seas.

I heard the front door swing open. I couldn't recall locking it. I glanced at my watch: not fifteen minutes had passed since

Arthur's departure. I slid off the crate and crouched low on the floor, reaching for the revolver. I crawled to the half-open door to the showroom and peeked out. Two men, one of them huge, were silhouetted in the display lights. The smaller man stepped behind the counter. I raised up slowly, pointed the revolver straight out with my left hand and gripped my wrist with the right, took a deep breath, and kicked the door open.

"Hands up!"

"Put that thing away, sport, or use it on Art Fletcher, not me," said the smaller man. "Don't worry, Hal, he won't bite."

"Eli! Goodness, don't scare me like that! What are you doing here?"

"Aren't I welcome?"

"Of course you are, it's as much your store as anyone else's."

"No, that's my store across the street," said Eli, jerking his thumb in that direction. "I broke that account sixteen years ago, Hal. We were the first New York jewelers in Jordan Marsh. They've carried us ever since. Until this year. Until this store. Where the hell are the lights, Jackie?"

I walked by Chase and snapped the switch. Eli looked about.

"Nice layout. Did you design it?"

"With Arthur."

Eli paced the length of the showroom, looking at his watch. "Twelve paces in eight and a half seconds," he said. "That's for a man of average build on a foggy Wednesday evening in October. Write that down and see if it matches Arthur's calculations. I need some cash. Where's the safe?"

"In the storeroom, but I don't have the combination. What do you need money for?"

"Expenses."

"There probably isn't much, just the workmen's salaries."

"How much would that be?"

"How much do you need?"

"Hal?"

"Twelve hundred," said Chase.

"And fifteen hundred for me. Twenty-seven hundred."

"There wouldn't be anything near that in the safe," I said.

"I'll take checks made to cash."

"They'd need Arthur's signature."

"Shit. Don't you have a checkbook?"

"There's nothing near that in my account, Eli."

"I'll have it for you in New York."

"I'm not going back to New York, Eli. I'm staying here until the opening, and then I'm sailing for Europe."

"I'll deposit the money in your account. Damn it, Jackie, write the checks! Or do I have to take some of the merchandise?"

I put the revolver on the counter and reached into my pocket for my checkbook. Eli took some slips from his pocket. "One for five hundred, two for three hundred, two for two hundred," he said. "Hal?"

"One big one is all I need," said Chase.

"Eli, could you explain to me how you could lose that kind of money on a tie game?" I said.

"Who said I lost anything? I told you this is for expenses." At this, Chase laughed.

"Make certain you cover these in New York," I said as I handed Eli the checks. "I can't foul my credit while I'm abroad."

"Don't give it a thought, sport. I'll take care of everything." Eli handed a check to Chase, who folded it into his hat.

"Thanks, kid," said the player. "Do you need any passes for tomorrow's game?"

"I said I'm not going back to New York."

"Neither are we," said Chase. "Didn't you hear the Commission's decision? They're playing here again tomorrow."

It was a game of ghosts, everyone said. The noonday sun dissipated the morning fog, but it began to billow out of the fens as play proceeded and by mid-game the field looked like the top of a moving cloud. To see the ball clearly the fielders crept closer and closer to the plate; the dimensions of play shrank to those of a sandlot game. The ball grew heavy from the sodden grass and needed to be replaced every other inning. The Giants managed a pair of runs, and Marquard, in midseason form, didn't allow a man past second until the ninth, when darkness and fog confused matters impossibly. The Red Sox pushed a run over the plate and stayed alive on a two-out error by Merkle. On the final play everyone heard the bat's sharp contact with the ball but no one actually saw the catch in the outfield; Boston argued furiously that the drive had landed safely, but Devore came out of the mists with a baseball in his glove and the Giants had a victory, tying the Series.

Boston seemed nearly empty on Friday; the Royal Rooters were back in New York to watch Tesreau and Wood rematched. Boys barked reports into the store while glaziers finished the counters. Arthur calculated that we received twenty minutes of work over the two hours of game play. Tesreau was good, but Wood was superlative, and the Giants' chances ended when Fletcher was out at the plate in a violent collision. The young clerk wouldn't part with the shamrock for fifty dollars.

On Saturday the pull to Fenway Park was unendurable. I searched out the oldest and most thoroughly orthodox synagogue in the city and buried myself there until the sabbath ended at sundown. When I emerged the newsies were shouting the result: Boston 2, New York 1. I bought a paper and read that the deciding run was the result of Doyle's error; Mathewson had retired the last seventeen men he'd faced. With any support in the field he'd have two victories rather than a tie and a defeat, and Boston, not New York, would be a game from elimination.

Nothing moved in Boston on Sunday, and Arthur and I passed the day working on the stock with polish and rags. On Monday the painters took their turn in the store. In the afternoon I rode the Boston subway—a child's Christmas toy compared to New York's—to the rail station to await our family contingent. They detrained at dusk; Edith carried Matthias, not the youngest in the pack of relations that crowded the platform. Ruby's whole brood was there, and so was his brother Saul's; my brother Carl had his bunch along. Lastly Uncle Sid alighted, a patriarchal smile on his face. We took a caravan of taxis to the new Copley Plaza Hotel, where Ruby's oldest boy showed me a sheath of play-by-play reports he'd purchased from a young entrepreneur aboard the train. The fellow had dashed to the telegraph office at every stop and printed the penny sheets between stations; he had the final score from New York—Giants five, Red Sox two—before the Boston press put it on the street.

There was a grand dinner that evening in a set of private rooms at the Copley, and at dessert Arthur read out the order of march for Wednesday morning. He allowed fifteen minutes for an address by Mayor Fitzgerald, and I interrupted to say that by all reports Honey Fitz was incapable of such abbreviation; the schedule was amended.

"Will Matty really be at the opening?" asked Ruby's boy Bill.

"That will be the finale of the event," said Arthur. "He'll present an award to your Uncle Jack."

"What if the Series ends tomorrow?" Ruby asked; the children booed the very thought, and Bill assured him that Mathewson would prevent that; Ruby amended his question. "If for some reason Mathewson doesn't pitch tomorrow and the Series ends, will he stay over?"

"I have Frost's assurance that he will," said Arthur.

"Are there tickets for the game?" asked another of the youngsters.

"That can be arranged."

"Where's Uncle Eli?"

"He was at the baseball today," said Uncle Sid. "He's coming late tonight."

"You're all free tomorrow until dinner," said Arthur. "They'll be laying carpet at the store, but you're welcome to visit at any time. Are there any other questions?"

"What's the cost of the carpeting?" asked Saul, head of accounts, and the affair broke up in laughter. Uncle Sid remained at his presiding chair, and Carl whispered that I ought to stay. I asked Edith to take Matthias up to bed and waited for the room to clear. Finally Arthur closed the door and took a chair at the foot of the long table, opposite Uncle Sid. The old man handed Carl an envelope, which Carl carried to me. It contained a wealth of twenty-dollar bills in mint condition.

"I think maybe you need cash," said Uncle Sid. "By the bank on Friday you overdraw two thousand dollars and some. I covered it, you shouldn't worry. You gave money to Eli? Checks for cash?"

"He told me he needed it for expenses," I said.

"Expenses!" Uncle Sid unwrapped a cigar. "Already he could buy a whole team of baseball with his expenses."

"It's a nice euphemism," said Arthur from his distant perch.

"He has no other excitement," I said.

"I refuse to be so blamed!" Arthur shouted, slamming the table. "I've protected you all from his shenanigans. I've done the work of two men, three! I've readied this store, I've taken leases on three others, I've doubled sales everywhere in New England and covered Eli's flanks in other territories, and all the while he's sought to bring the firm to ruin rather than see me succeed!"

"Don't be melodramatic," said Carl, placing a beaker of brandy in front of Arthur. "He doesn't see you as a nemesis, just a nuisance. You ought to return the estimation. He's never opposed any plan of yours."

"How could he? He must see the inevitable necessity. But it's gone beyond the point of supporting a wayward brother. It threatens to run to unmanageable sums. Stopping his account at the bank didn't work, he merely turned to Jack. One way or another he'll continue until he takes us all down. I've asked you if there are legal remedies, and you say you have a proposal. What is it?"

Carl took a folded packet from his breast pocket. "This is an instrument which would enact certain provisions of Uncle Sid's will in his lifetime," he said. "It establishes family ownership of the firm."

"But the family already owns the firm," I said.

"Not in the eyes of the law. Legally, Uncle Sid is the sole owner and officer of Pincus—of Collegiate Jewelers. This instrument alters that fact. It calls for a division of shares among the family while Uncle Sid retains voting rights. It also creates bylaws for the corporation and calls for a board of directors. Those of us who work with the firm will sit on the board, and we'll receive voting rights when Uncle Sid . . ."

"Dies," said Uncle Sid.

"Or retires," said Carl.

"Dies," the old man repeated.

Carl sipped some brandy. "The board will vote dividends every year," he said. "Board members will receive stipends, and of course you'll all have your salaries. The dividends, I should say, will be generous; that will reduce Uncle Sid's eventual estate, and avoid death duties. Furthermore——"

"What has this to do with Eli?" said Arthur. "How would this restrain his incorrigible behavior?"

"Arthur, this is not easy for me!" said Carl. "The man put me through school! I'm trying to put something together that won't be a slap in the face."

"He needs more than that," said Arthur. "We have to——"

"Avram!" Uncle Sid had been listening with his hands folded over his forehead. Now he put them on the table before him. "Long time ago," he said, "when you don't got your first long pair of pants yet, when you, Yakov, you want to go off playing baseball in Australia, and when you, lawyer, just out of school, you think the family business is *verdreck* and got other fish to eat, all that time where was Eli? On the road. Trains, hotels, sometimes not even a hotel, a bench in a station some nights. Comes home to what? Nothing. Who marries a man on the road all year long? You spend all your life on the road, you see what it makes a man. But you see, too, what it makes us! Because when he comes off the road, what's he got? Orders, big orders. We start a little shop, now we got a factory bigger than the whole *shteitel* I was born. Who did that? Eli did that.

"Now, Yakov comes in. Does he help Eli? No, he's gotta draw. Okay, good for you, we need you to draw, but Eli's still alone. Then you, Avram, you come in, but do you want to help Eli? No, you gotta do everything different, have our own stores, change everything. Okay, that's the way the business grows, I got a responsibility to do it, but Eli's still alone. And now you want maybe I should threaten him? Fire him? Fire your brother?"

The old man leaned back and put a match to his cigar. "Now, I don't like what Eli does on the road. I don't like the *courvahs,* I don't like the *vonces* he hangs around with, I don't like the gambling, and I don't like the bank in my office *alle Montag unt Donnerstag,* he overdraws the account again. I say, why he does this? Does he want to ruin the family, like Avram thinks? No, he doesn't want that. He thinks it's like a faucet, with the water runs however much you want. He don't remember a *shteitel* with a well, you get one bucket because everyone needs. He's in America, where the faucets run forever.

"So, what do we do? We make plain what's his. I know Eli, he don't take from family, he don't take what's not his. Give him, tell him this is yours, he'll live on it."

"That's the point," said Carl. "It will be made clear to everyone in the family, not just Eli but everyone, that barring a catastrophe the dividends, and in Eli's case the salary and stipend, are the limit of the family's support. And a catastrophe does not mean a gambling debt. It means a fire or an illness or some other act of God."

"He'll end up losing his stock," said Arthur, "and we'll have some Irish gangster sitting on our board of directors."

"Can't happen," said Carl. "The instrument prevents the sale of stock to any outsider without unanimous approval of the board. The board itself, or any member or combination of members, has first rights to the shares."

"Oh, that's good," said Arthur. "That's very good. Yes." He sipped his brandy. "I like that very much."

"I'm going to explain this to the whole family tomorrow after dinner," said Carl. "I'm very concerned that Eli doesn't take any of it personally. He mustn't suspect that his behavior prompted the plan."

"Who else knows the truth?" I asked.

"No one outside this room," said Carl. "You'd not have known but for the checks you wrote."

I shook my head. "I can't say I like it," I said. "The formality makes us something less than a family firm."

"Not something less," said Arthur. "Something more."

"There comes a point when formality has its advantages," said Carl.

"Come here, Yakov," said Uncle Sid. "And you, leave us alone."

My brothers took a last draught of brandy; Carl collected the packet of papers; Arthur closed the door softly behind them. I sat down next to Uncle Sid, and he took my hand in both of his.

"Tell you true, Yakov, I don't much like it no more than you," he said. "But so big we're getting! You think I ever dreamed we get five hundred stores to carry our line, like the Matson deal got us? And now we open our own store here, next year three or four more, soon more and more. So big! Yakov, failure got no problems. If we fail we got no problem with Eli, no problem with anything. We just curse and cry and go to the next thing. But success! Success got a million problems, new ones all the time. So? We do what we got to do. For the family."

Edith had the clever notion that we might give Matthias a start on his sea legs with a day trip by lighter to Cape Cod, and it proved a popular outing; nearly a dozen of us shivered in the wind and spray. When we came to dock late Tuesday afternoon Ruby and the older boys were waiting at the pier, the youngsters waving their Giant pennants gleefully. "We won!" they shouted. "We killed 'em, eleven-four! The Series is tied!"

"It was a slaughter!" young Bill enthused. "Wood didn't last to the second inning! Doyle hit a home run!"

"Mathewson must have coasted," I said. His season was done, the victory crowning it; I looked forward to the morning and its ceremony.

"No, it was Tesreau, not Matty!" Bill said. "McGraw's a genius! He's got Matty ready for tomorrow, and Wood's all used up. The Series is in the bag!"

"Damnedest thing I ever saw," Ruby added. "Six Giant runs in the first inning, and the Royal Rooters turned on their own fellows like a lynch mob. The park was nearly empty by the sixth. They've given up. No one believes they can beat Matty again tomorrow."

"He won't come to the store tomorrow morning then, will he?" I said. "Not if he's pitching the final game."

"Arthur will have apoplexy," said Ruby, "and on top of it all Carl's called some kind of meeting to explain a reorganization of the business. It should be quite an evening."

I had an escape: it was the eve of our wedding anniversary. We paid our respects at the family dinner, saw Eli amazing the children with card tricks and Arthur worrying over his schedule, and then we stepped out. Uncle Sid had arranged the hire of a gorgeous carriage drawn by a matched team of stallions, an all but extinct luxury amid the increase of automotive traffic. Edith and I attended the symphony, then shared a late supper at the Locke-Ober Cafe on Winter Street. We arrived back at the Copley in great pomp and high spirits, well after midnight.

"There's a Mister Fullerton waiting for you, sir," the night clerk informed me.

"Hugh Fullerton? The baseball writer?"

"Yes, sir, here's his card. He's waiting in the lower lobby."

I explained the matter to Edith and then sought out Fullerton, remembering the rimless glasses and western hat that had distinguished him in the locker room after the Merkle game. When he stood to shake my hand he towered over me, nearly Mathewson's size. "This place is thick with Kapps, you're the fifth one I've hailed tonight," he said. "There's a quiet public house a few blocks away that congenially ignores the curfew. Come with me."

"Must I, Mister Fullerton? It's my anniversary, and tomorrow's an important day, we're opening——"

"I know all about that. It's why I want to talk to you."

"Really, Mister Fullerton, if it's a story you're after it must wait until the morning. I was about to retire."

"I'm not on a story. I've spent the evening with Matty. If you have any feeling for him at all, you'll come with me. Please. It's not far."

The pub was as dark as the midnight streets and no more crowded. Fullerton led me to a table against the wall, removed his hat, and ordered a stein of dark beer. When it arrived he blew the foam from the top of the stein.

"These eastern beers are nothing like Chicago's home brew," he said. "Sure you don't want anything?"

"Quite sure. Now what's all this about, Mister Fullerton?"

"It's about your friend Christy Mathewson. I think you ought to know that he's in an awful state. It's been a terrible week for him—the blow-up with McGraw, the incident on the train——"

"I know nothing of any of that."

"Don't you? I'm surprised. Matty and McGraw had a set-to about the assignments for the Series. Matty expected to start the first game. He specifically wanted to go head-to-head against Joe Wood. Worst argument they've ever had, worse than when McGraw pinch-hit for him in the playoff in 'aught-eight. I'm sure you remember that. Then on the way up here last Tuesday some drunk broke into Matty's compartment in the middle of the night. He only wanted to wish Matty well, but a private dick pulled a pistol and there was—well, an incident. McGraw kicked the poor fellow in the privates while the dick had him pinned." Fullerton sipped the beer. "And on top of it all there are these punk Royal Rooters and the hell they've given him during the game, the abuse of Jane and his kid. Haven't you heard all that?"

"I haven't seen any of the Series games. In fact, I haven't been to the park all year."

"You haven't? You?"

I tried to explain my year-long abstinence in terms of artistic renewal; Fullerton casually accepted the excuse.

"Well, since you haven't seen it, let me inform you that the team has played like a passel of clowns behind him," he said. "When he lost on Saturday he was very despondent. He told a reporter he might never pitch another Series game, and though I'm sure he meant that he might never have another opportunity it got back to McGraw that he'd refuse another assignment. Mac lit into him like a mongrel dog. Called him a quitter and I don't know what else. There's no talking to McGraw when he's in one of those black moods. They didn't speak until after today's game, when Mac said Matty would be on the lineup card tomorrow and if he didn't take the mound he'd have to tell the world the reason why.

"I had a call from Matty about eight o'clock tonight. We're old chess partners, I thought he wanted some relaxation. We began to play, but his mind was somewhere else. I've never seen him make worse blunders. Finally it all started to come out. What could I tell him? I said he was putting the pressure on himself, and that no matter how it felt it was a good sign, he was getting ready to pitch the game of his life."

"Mathewson performs best under pressure," I said. "Do you know that the Dutch jeweler who cut the Kimberley Diamond was so appalled by the responsibility that he fainted when he struck the blow? But he struck it perfectly."

"Is that why you've decided to add to the pressure?"

"What do you mean?"

Fullerton took a folded telegram from his notebook and dropped it onto the table in front of me. It read:

CHRISTOPHER MATHEWSON
NEW YORK GIANTS BASEBALL CLUB
SAINT JAMES HOTEL
BOSTON, MASSACHUSETTS

WIRE RECEIVED THIS INSTANT FROM COLLEGIATE JEWELERS QUOTE
IMPERATIVE MATHEWSON PARTICIPATE IN CEREMONIES
TOMORROW MORNING FUTURE RELATIONSHIP DEMANDS IT
UNQUOTE DECISION ENTIRELY YOUR OWN BUT LAST YEARS
INCOME FROM COLLEGIATE NEARLY FIVE THOUSAND DOLLARS
FROST

I felt ill. "I had nothing to do with this," I said. "I have a brother . . . well, I'm sure he sent the wire. I don't imagine he thought that Frost would forward it to Mathewson. Arthur believes in more subtle forms of manipulation."

"You have quite a set of brothers. I know Kappy."

"Everyone knows Kappy. What did Mathewson say about this?"

"He talked about you and your art. He respects your work enormously."

"So he told me, the only time we talked at any length."

"The only time? What do you mean, the only time?"

"In September of 'aught-five, at dinner at my father-in-law's club. Other than that we've only met in passing. There's been some correspondence, very polite. I last saw him, when? Four years ago, in the Giant clubhouse."

"You can't be serious." Fullerton's stein began to tremble, and he held it on the table with both hands.

It wasn't unusual for people to assume that I was close to Mathewson; I was, after all, the designer of the line that bore his name and endorsement, and the rings I'd fashioned for his own hand were hardly secrets. "You mustn't think that my work reflects a personal intimacy," I said. "I've never claimed——"

"No, no, it's not that," said Fullerton. "It's what Matty himself said about you. He showed me the rings you've done for him, yes, but he also said that you'd seen him pitch every important game of his career, that he has a gift from you on the fifteenth of July every year, that you followed him on the road, that you were . . . what were his words?"

"His greatest fan?"

"Nothing so mundane as that. No, it had a special connotation, something . . . the celebrant of his works. That's what it was."

"The celebrant of his works!"

"Yes. He certainly gave me the impression that you were intimate. And then he dictated a note to you."

"To me!"

"Yes. He'd have written it himself, but his hands were shaking so badly——"

"No!"

"Indeed they were. I have it written down." Fullerton took a notebook from his breast pocket. "Here it is. 'My dear Master Kapp, I ask you earnestly to excuse me from the obligations of the morning. My mind is not free to pay you the tribute you deserve so long as it is fixed upon this other task that lies ahead. I would despair if this adversely affects the bond between us. I pray that I shall give you cause for celebration.' "

I took the notebook and read the words. "So he sent you to see me," I said.

"No, he asked me to copy out the note and see to its delivery. I decided on my own to bring it personally. Kapp, is it true? You've not spoken at length for seven years?"

"Not at all, except to shake his hand in the clubhouse, and not that since 'aught-eight."

"And before that?"

"I was in his apartment once in 'aught-four, with McGraw and Eli. Kappy. It was unplanned."

"He gave me a totally different picture. I thought you were very close."

"I've never even been alone with him," I said. "He certainly has a hold on my imagination, that's been the case since I first saw him pitch. July fifteenth, nineteen and one."

"The no-hitter in St. Louis. I know the ring."

"Still, I've never stopped to consider what I might be in his mind."

" 'The celebrant of his works,' " said Fullerton. "As are we all. Have you ever considered what he is to himself? What it's like to be Christy Mathewson? Imagine it. You know perhaps five hundred people by name, but fifty million know you. You make no more than ordinary demands upon people; you don't insist that the sandwich you order for lunch be the most marvelous sandwich ever made, or that the bootblack's shine dazzle the blind, yet the sandwich-maker and the bootblack and millions like them expect the superhuman from you, and finally they'll accept nothing less. Expectation becomes demand, and it extends to everyone and everything. You hear the crowd groan if you give up a single hit; they expect a no-hit game. Give up a run and people say you're off your game. Even your teammates turn to you to save them after they foul up the simplest plays. The writers make you the standard of excellence, and if a rival wins nineteen games in a row you're expected to win twenty. The world makes you a god and hates you for being human, and if you plead for understanding it hates you all the more. Heroes are never forgiven their success, still less their failure."

"Not me. Never me."

"You and me and my brethren of the press corps, more than anyone. All the celebrants of his works. We make the greatest demands. Every time he pitches I find myself hoping for the most extraordinary achievement, for my immortality lies in his. No one will ever dig my columns out of the morgue for the sake of my literary genius. It will be to read about Christy Mathewson, or Smokey Joe Wood, or Tris Speaker, or any of these demigods we create and celebrate, but Matty most of all, for he's the master. I want him to throw a no-hit game tomorrow, not for his sake but for mine. And don't you want the same, so you can cover him in glory? We're the worms that eat at the bodies of the great.

"And you're the high priest, Kapp, the celebrant-in-chief. He demands that you equal in fashioned stone anything he may do on the field. Do you see that challenge? And do you hope, deep within, that he'll fail and relieve you of the burden? Is that why you stayed away this year, not in fear that he'd fail and leave you without inspiration, but terrified that he'd succeed and leave you to match it in your work?"

"I can't think it," I said.

Fullerton put the notebook in his pocket and came out with a small envelope that had my name written on it.

"Matty thought that with the change in schedule you might need a pass for tomorrow's game," he said. "I think you ought to use it. 'I pray that I shall give you cause for celebration.' I mistook his use of the word, thinking of dancing in the streets. But he addressed it to a celebrant, didn't he?"

Fullerton put on his hat. "Matty told me you were once a pitcher. I suspect that your work is infused with the wish that you were he. You're not alone. Inside every sportswriter there's a frustrated athlete, according to the old saw. Why not? The same thing is inside every fan, or anyone who ever picked up a bat and ball. But Kapp, you ought to thank God that your arm went bum. It might be you in Gethsemane tonight."

Arthur spent the morning in his own agony. It was the great moment in his life, and it dazed him to discover that the baseball community placed a lesser evaluation on the affair. He thought that bands and bunting and speeches galore would draw the national sporting press, but he overlooked the single element that would have guaranteed coverage: free food for the writers. The family outnumbered the journalists by an embarrassing margin. Neither Mathewson, who conveyed his formal regrets through Frost, nor any other player of either team attended; they'd rather contest for the rings than gaze upon them. No Royal Rooters paraded by. Disgusted by the one-sided defeats that had brought the Series to a tie and dismayed that the mighty Mathewson was rested and ready, they'd torn their sashes and littered Kenmore Square with their top hats.

Still, the ceremony was no mean occasion. All three members of the National Commission were there, maneuvering for position in the photographs: Ban Johnson, the beast of McGraw's imagination; Garry Herrmann of Cincinnati, the Commission chairman; and the third man, National League president Thomas Lynch, who'd succeeded Harry Pulliam in 1909 when that unfortunate man had stilled McGraw's frothing anger by blowing his own brains out in his rooms at the New York Athletic Club. "I didn't think a bullet to the head could hurt him," was McGraw's epitaph. There was a smattering of other officials and a few of Boston's alumni: Honest John Morrill, who'd managed the Nationals in the 'Eighties, and the Wright brothers, George, Sam, and Harry, from the antediluvian age.

It was Honey Fitz who saved the occasion. The Mayor, and the scream of sirens that announced his passage, was a magnet to Boston's Irish, and if few who stampeded after his emerald banner could afford our prices the crowd made a rambunctious scene for the motion picture camera that Arthur had hired. There was also an oversized contraption to record the speeches, and it intrigued the Mayor.

"An entire oration preserved on this single waxen disc!" said Honey Fitz. "What a remarkable application of technology!"

"Not an entire speech, your honor," said the man who operated the device. "Each disc has a ten-minute capacity."

The Mayor made a swift, silent count. "Better get a dozen more," he said.

A genial spark flashed between Honey Fitz and Uncle Sid, who were a matched pair of size, shape, and years, and in scarcely a minute the Mayor had a history of the family's emigration, travail, and triumph. This the Mayor took for his theme, and before he'd finished his oration he'd made of the Jordan Marsh store the Established Church of England and of our shop the meanest synagogue of middle Europe. When he was done Uncle Sid embraced him and pledged him fealty. "You can use the store as a voting address," the Mayor replied.

Arthur was tied to the notion that an active player must present my award, and the only available candidate was Hal Chase. The honor meant nothing to me, and I accepted Chase as its donor for Eli's sake; I had a kindly thought for the player as I watched him renew his acquaintance with Matthias. Chase spoke briefly and effectively in the pose of one who had long coveted a championship ring but whom fate had disappointed. I said, "Thank you," and turned away, and the audience, no doubt grateful to be spared another speech, applauded enthusiastically. The band played

"Johnny, Fill Up the Bowl"; Honey Fitz unlocked the front door with a golden key and made the first purchase, a modest brooch for his eldest daughter; he insisted on paying full price, and Saul's infant son, guided by his father's hand, rang up the sale. The crush inside the store threatened the Mathewson display, and I held it aright, strangling the model of the pitcher with my hands. Then the Mayor was gone, and with him the crowds; the workmen set to dismantling the sidewalk platform; the band dispersed. The laughing children played at clerks and customers while their elders stared about with expressions of hope and pride and apprehension. Uncle Sid invoked a prayer, and the children fell quiet. He finished with a muttered "amen" and a shout of *mazel tov!*

Amid the chatter I sought out Carl and asked how the discussion of the firm's reorganization had gone the night before.

"Eli's wiser then we know," he answered. "He asked if the plan were Arthur's idea. I reminded him that I'm the family attorney. After that it was just 'where do I sign?' Happy anniversary, by the way. You two looked like the King and Queen of Sheba in that coach last night. And bon voyage! I might not see you before you sail."

"No, I'll bc at the game with you."

"Really! Breaking your fast, are you?"

"I am," I said. "I feel as if I've been called to witness."

Fenway Park, a late model in the outbreak of stadium construction that excited both leagues in their prosperity, seemed a marvel of architectural simplicity: a crater dug in the ground, its bottom flattened for a playing field and rows of seats set on the slopes. Yet the hole seemed carelessly, even drunkenly cut, and the field of play was violently asymmetric: right field was an enormous pasture, left field hardly existed. The high wall in left loomed so closely over the infield that an outfielder there seemed a superfluity. At the base of the wall a steep embankment sloped ten feet downward, and during the season that hillside was part of the field of play; it was called "Duffy's Cliff" in honor of Lewis, who patrolled it like an Apache scout. For the Series a set of temporary bleachers sat on the incline, but they were empty this day, as were nearly half the seats in the park. The New York crowd, in a muddle over hotel and train reservations since the tie game fouled all plans, had departed the night before, swearing to greet the victory train at Grand Central. The Royal Rooters, despondent and disenchanted, and in a snit that no free passes had been issued to their numbers, stayed away.

I had the passes from Fullerton in my pocket, but I chose to sit with the family, which spread out over an uncrowded section of the right field corner. Ruby's boy Bill, loaded down with books of statistics he'd compiled over the season, stayed beside me. The children glimpsed Eli in a faraway box with Hal Chase and beckoned to him; to their delight he answered the call. We greeted him, we fifteen brothers, cousins, nephews, and laughed when he shouted that we had a team to equal either on the field.

I watched for Mathewson as a child watches a masquer, at once frightened and compelled to discover if Fullerton's report was honest. I saw a huge, broad-backed figure spring out of the Giant dugout, alive with youth, and I came to my feet. Young Bill looked at me strangely. "That's Tesreau," he said, and I froze, then pleaded the distance to home as my excuse.

"There's Matty," said Bill, and there he was, a mechanical man who set to his warm-ups with the dull application of a tarrier pounding track. McGraw kept his distance, moving behind Meyers to judge Mathewson's pitches like an umpire and then slipping back into the dugout.

I reached into my pocket and touched the envelope from Fullerton. "Would you like to move down closer, Bill?" I asked, but then Eli's hand was on my shoulder.

"How does he look to you, sport?"

"Tired," I said.

"But is he ready?"

"He's always ready," said Bill. "He's Christy Mathewson!"

"Billy, have you ever heard about the first time your old uncles saw that man pitch?"

"The no-hitter in St. Louis? You told me the story the first time you took me to the Polo Grounds, Uncle Eli."

"And did I tell you what it cost me?"

"You bet against him?"

"I wish I had." Eli laughed. "I don't think you're old enough for the whole story, Billy, but if you ever get out to St. Louis I've got some red-hot numbers for you. My, oh my. Have there ever been so many of us at a ballgame? Put that in your records, Billy. October sixteenth, nineteen hundred and twelve: most members of the Kapinski-Pincus clan at a single game, sixteen. Note that your Uncle Arthur and your grandfather did not attend, and that it's the first game all year I've seen with your Uncle Jackie." Eli took a nickel from his vest pocket and spun it in his fingers. "Let's make it interesting, sport," he said. "Take either side, and I'll cover it."

"No bet, Eli."

"I'll take the Giants!" said Bill.

"You're covered." Eli flipped the coin over the boy's head to me. "Hold the stakes, sport. Come on, Billy, ante up."

"I've got a quarter," the youngster said as he dug into his change purse, but Eli shook his head. "I'm on a tight budget, Billy," he said, looking at me. "A nickel's as much as I can handle. Now let's watch the game."

I watched Mathewson. It seemed he'd measured the task against his store of energy and resolved against a single extraneous motion. He performed none of those characteristic gestures so familiar for a decade, the pull at the belt, the tug at the bill of his cap. He had the strength to bend and stride and throw, but if Meyers' return toss was off line, as twice it was in the early going, Mathewson wouldn't reach for it, and play had to wait until Snodgrass returned the ball from center field. He spoke to neither teammate nor foe—a stark contrast to McGraw, perched on the dugout's top step to direct and encourage every Giant save Mathewson and to bait every Boston batter. The man on the mound was never so alone.

His pitches were of the purest form: strikes at the edges of the plate, teasing off-speed deliveries, magical drop curves. He struck out two men in the first inning; he struck out Speaker to close the third; he struck out Stahl to end the fourth. He threw three pitches in the fifth; each became a soft fly ball. By then he was protecting a one-run lead; he'd nearly doubled home two more runs, but Speaker had caught his drive at the limit of his range.

In the sixth Boston moved a runner to third and tried to squeeze him home with the tying run; Mathewson prevented it with a pitch that flattened the batter, and then he sprang to the baseline to take Meyers' peg and tag the retreating runner. A moment later he began the seventh with a base hit, and when Devore bunted he tore into second base and upended the shortstop in a violent out. They disentangled, and Mathewson sat on the base for a long moment until the umpire urged him back to the Giant bench. He walked the way like a soldier retiring from furious battle, and his teammates gave him a wide berth on the bench. Silent and still, he watched the rally die.

In the Boston seventh Stahl blooped a hit with one out, and Mathewson missed narrowly on four pitches to Wagner. Here was danger, and a rising hope in the cries of the home fans, but Cady popped weakly to shortstop. The crowd groaned. Our own section was dead silent, and Bill had a death grip on his notebooks. Stahl came off second base and pointed at his dugout, and a small player emerged, swinging a bat left-handed.

"Swede Henriksen," Bill whispered. "He's a good hitter."

Boston's pitcher slapped his replacement on the shoulder and went back to his bench; the young pinch-hitter took some practice swings, walked slowly to the plate, stopped at the edge of the batter's box, and swung some more. Finally the umpire had to order him into the box. Having swung thirty times at phantoms, he froze at Mathewson's first pitch, a strike at the knees. On the second his hips and shoulders came around, but not his bat; strike two.

"Has he seen Matty before?" Eli asked.

"It's his first time up in the Series," Bill answered.

"Fadeaway," I said. "Watch."

There it was, spinning off the corner. Henriksen swung late and nicked it with the very end of his bat, and the ball sailed softly over Herzog's late-rising glove at third—an accidental hit, but a double to tie the game and leave runners at second and third. A white wave in the Giant dugout caught my eye: McGraw, kicking a towel the length of the bench. On the mound a stoic Mathewson stood upright with his hands at his chest, squeezing the ball between his bare hand and the palm of his glove. Henriksen, at second, was fighting back an idiotic grin. Mathewson threw to Hooper, who hit it well—a thunderclap compared to Henriksen's effort—but straight at Snodgrass in center, and that fielder, surest of any Giant, gloved it easily.

The Boston faithful applauded the team as it took the field for the eighth and it grew to an ovation for the pitcher Stahl now called upon. We watched the new man walk across the outfield, and Bill gulped. "Smokey Joe Wood," he said in the tone of a deathbed mourner. "Smokey Joe Wood in relief."

"I'll make it a quarter on Boston now," said Eli, and when young Bill failed to respond he added, "at two to one."

"I'll take that," I said.

"God of light! A new day is dawning. Did they give you an allowance too, sport?"

I ignored Eli and reassured my nephew. "Mathewson's wanted to go head-to-head against Wood all week," I said. "Now that he has it he won't let go until he's won." I felt in my heart that it was true, that all the elements of chance, including Wood's first-inning failure the previous day, had fallen together to create this ultimate challenge. I saw Mathewson move from the bench to the dugout steps to study Wood in his warm-ups. He was near the fire.

Wood's inning was all fastballs and quick results: two ground outs, a sharp single by Herzog—if only the man could field!—and another grounder. There was new life in Mathewson's step as he came to the mound. His first pitch came back at him, inches off the ground; he kicked at it, and the ball caromed off his calf and landed at the feet of the surprised Herzog, who had time to measure his throw and put the runner out at first. As Mathewson moved against Speaker he stumbled and aborted the pitch. McGraw leaped out of the dugout, but Mathewson turned his back abruptly to stare into the outfield. McGraw stopped short, shrugged, and retreated. Mathewson fell behind Speaker, struggled back to even the count, and retired him. Duffy Lewis did no better than Speaker. The game, the season moved to the ninth.

The family, once scattered over a stretch of seats, had drawn close about, and Ruby took out his timepiece. "When do you sail?" he asked me.

"Sail? Oh, at midnight."

"It'll be dark before long," he said.

"They won't call this game until it's pitch black," said Eli.

A fly ball down the line sliced toward us, and Harry Hooper drifted over. The youngsters screamed to disrupt his play, but he caught it directly in front of us and, after returning the ball to the infield, he flashed a smile at the children. Bill dropped his pencil.

"Why, he smiled at me!" he said. "Harry Hooper smiled at me! Just like a real person!" He checked his books, stood, and cupped his hands around his mouth. "Hey, Hooper!" he shouted. "You're batting three hundred for the Series!"

The fielder raised his glove.

"He waved at me!" said Bill.

"Just like a real person," Ruby laughed.

Mathewson came to bat, but he had nothing left for hitting; after three perfunctory swings at Wood's fastball he returned to his dugout isolation. Wood passed Devore and retired Doyle on a grounder. The Red Sox came in for the bottom of the ninth, and Hooper winked in our direction as he trotted by.

"Wow!" said Bill. "What a shame he's on the wrong team."

Mathewson went to work and retired Gardener. Stahl followed with a fly ball to left; it plunked off the neighborly wall inches over Murray's glove, and Stahl ended at second base. When the ball came back to Mathewson he hitched his belt. He worked the outside corner to Wagner and got a fly ball to the generous area of right field where Devore caught it for the second out. Meyers stood up behind the plate and stretched out his glove; he wanted four intentional balls. With a sharp movement of his hand Mathewson ordered him to squat.

"Why won't he walk Cady and pitch to Wood?" asked Bill.

"It wouldn't be Wood," said Eli. "They'd pinch-hit for Wood, he'd be out of the game. You're right, sport. Matty must really want Wood head-to-head. He's keeping him in the lineup."

Mathewson threw two balls high to Cady, then a strike, then another. When he took Meyers' throw he hitched his belt again and straightened his cap. He threw once more, and Cady popped the ball to shallow left field; Murray trotted in, tapped his glove, and caught it.

"Extra innings!" said Bill. "There's never been an extra-inning game to end the World's Series."

"How about the Temple Cup?" said Eli.

"The what?"

"Ancient history, kid. Never mind."

Smokey Joe began his third inning of work by retiring Snodgrass, and as Red Murray dug in the children chanted "We want a hit!" Murray tapped the plate and brought up his bat, Wood kicked high and fired, and the ball shot out to left-center field. Speaker went after it, and so did Lewis; they met at the base of Duffy's Cliff to watch the ball drop into the temporary bleachers. Murray slowed to a trot, and all of us were up and shouting, but the base umpire, like a traffic cop, held Murray at second.

"That's not fair!" the children complained. "It's a home run!"

"No, a grounds-rules double," said Ruby.

"And here's Bonehead Merkle at bat," Eli said, and the children moaned. If they knew nothing more about baseball than that Mathewson could pitch, Speaker could hit, and Ty Cobb could steal bases, they knew that Merkle was a clown and an ignoramus, marked by that single play which no later success could erase. Never mind that he'd led the club in home runs this year and runs batted in the year before, never mind his eight hits in the Series; he was Bonehead Merkle forever. Wood stretched, then stepped off the rubber and wiped his hand on his thigh. Herzog's fearsome bat was next, there was no walking Merkle. In the Giant dugout Mathewson stood with both hands gripping the roof, his eyes intent on his rival. Wood dusted his hand with resin and bent for his sign. The children raised another chant, and boos rolled over us from the Boston crowd. Wood missed with a fastball and kicked at the mound. The children clapped. Wood went through the whole exercise again and threw a fastball with full force. It bounced a yard short of the plate.

"He looks done, sport."

"He does that."

The children were too excited to clap now, and the Boston

crowd so quiet that we could hear Wood grunt as he delivered his next pitch. Merkle went after it, and it was a blur over Wood's shoulder, then a spiral of dust near second base, and finally a tiny white prowler on the outfield grass. Speaker, so close behind second, charged the ball. I took my eyes off the play to glance at Murray rounding third and saw, as Murray did, the frantic circling of McGraw's arm in the coacher's box. In the same instant I heard the sharp gasp of the Boston crowd; I looked back to center field to see the ball on the grass behind Speaker. Then I looked home: Herzog had both arms over his head, the signal to Murray to come in standing. Murray stomped on the plate and leaped into Herzog's arms, then ran at full speed to the dugout. The first man to greet him was Mathewson. With the whole of Boston dumb, with play routinely continuing on the field, Fenway Park observed two wild celebrations, the Giant team on its bench and our family in the far corner of the grandstand.

"Do you know, I was four years old the last time the Giants won the World's Series!" said Bill.

"It's not over yet," his father cautioned.

"It's in the bag! It's in the bag!"

"Looks that way," said Eli, his smile as broad as any. "Sport, why not give Billy his chips?"

I reached into my pocket for the nickels I was holding and felt the envelope from Fullerton. "Billy, how would you like to see the finish close up?" I asked, and the youngster looked to his father, who nodded. I handed Bill the coins and stepped into the aisle with my hand over his shoulder.

"Sport!" Eli flipped a quarter at me, and another. I caught them, took Eli's hand in mine, pressed the coins into his palm, and closed his hand around them.

"I love you," I said.

"Don't be silly," he said.

"World's champs! World's champs!" the children cried as I walked my charge down the aisle toward the infield. On our way we drew some unkind shouts from the locals who had marked our contingent. I showed my passes to an usher who led us down to the first row of boxes and chased a pair of Bostonians who'd poached on our seats. We were thirty feet from first base, and when Merkle trotted out to his position in the bottom of the tenth Bill stood and applauded.

"You're the hero, Merkle! You're the hero at last!"

"Sit down, punk," said a man behind us.

Mathewson threw his warm-up pitches, signaled Meyers he was ready, and turned to face the outfield while he rubbed the ball. He appeared to be studying the scoreboard high over the bleachers in right, the long row of zeroes punctuated by a *1* in the Giant third and another in the Boston seventh, and a lovely white *1* in the top of the tenth. A slight smile began on his face, but he wiped it away with a touch of his glove and turned to the plate.

Clyde Engel was there, pinch-hitting for Wood. He was a very big and very broad brute, a natural for Fenway Park, where the wall transformed any right-handed hitter of size into a long-ball threat. Mathewson fed him soft curves that fell over the plate like snowflakes. Engel missed one outright, took another, and hacked at a third to send it foul into the Giant dugout.

"I still can't believe Speaker butchered that play," the man behind us complained. "He has Murray out a mile at the plate."

"He makes that play ninety-nine times out of a—no, make that nine hundred and ninety-nine out of a thousand," a companion said. "Why does he foul it up now?"

"Dumb luck," said the first. I thought of Mathewson's seven years' famine of luck: illness in '06, unhappy veterans in '07, Merkle in '08, the endless rain in last year's Series, the error-marred performances this year. With three outs he could erase the frustration of those years and resurrect the glory of his youth and the joy of my own, with the Giants at the top of the baseball world and my family whole and unstrained. When he moved into his motion, so balletic, so familiar and reassuring, I felt myself in the mob around the old Polo Grounds clubhouse, watching Mathewson unfurl the championship banner of '05, and I sensed the same tears welling inside me.

The pitch was another curve. Engel swung mightily but produced only a lazy fly ball over shortstop. Fletcher turned and ran easily, pointing at the ball high above him to spot it for the outfielders, and Murray and Snodgrass, from their separate posts, trotted to the place where it must land. Murray was a bit closer, but the play belonged to the center fielder and Murray pulled up short as Snodgrass waved him away and caught the ball in his thick open glove and made to shovel it to Murray, and it dropped onto the grass at Snodgrass' feet.

Engel was at second base; Doyle caught Snodgrass' throw there and tossed it softly to Mathewson. The pitcher snatched it out of the air angrily and stood with his hands on his hips, staring at center field. Snodgrass looked at him, and Mathewson swung his gloved hand high in the air and brought it down hard in a gesture of utter disdain. He stomped back onto the mound and kicked up a clod of dirt. Meyers ran out to calm him, but he chased the catcher back with a motion as violent as his signal to Snodgrass. He stood at the rubber while Hooper took his stance, leaned for his sign, then backed off and turned to the outfield and kicked again. The crowd drove him on. He reached for the resin and slammed it back onto the dirt. He took off his cap and shook his head. His hair was dark with sweat. He toed the rubber, read another sign, took his stretch—not the usual full second, only a slight pause—and drove at the hitter.

Hooper was ready for a fastball; in such a state a pitcher would throw nothing else. He drove it to the deepest part of the park, where the bleachers jutted jaggedly into right-center field. Snodgrass ran with all the speed of his youth, reaching out at the last instant and catching the very bottom part of the ball in the web of his glove. It stuck there by some miracle or accident, and in three steps Snodgrass hit the wall. He spun, took two steps toward the infield, and threw. Engel tagged at second and ran for third; Doyle took Snodgrass' throw in short right field, began a motion and abruptly held up. In the immense noise Engel slid safely into third, Herzog stepping aside for him.

I looked for Mathewson behind third, the pitcher's place on such a play, but I couldn't find him there. To my astonishment he was still on the mound, staring at Snodgrass with the same angry mien. The center fielder's play had purchased no forgiveness from the pitcher. He dug at the rubber with his toe as Steve Yerkes came to bat. His first pitch was a hurried fastball, high; for his next, Meyers had to leap to catch it and hold the tying run at third. The Chief held the ball, but Mathewson demanded it back; he backed up to the rubber and delivered with hardly a motion at all. It missed the plate badly.

"He can't walk the winning run aboard!" said Bill. "Not with Speaker next!" I'd never seen him do it, never in a decade, but I'd never seen this appalling ferocity either. The next pitch was no better, and Yerkes trotted to first.

"Let's go back to the family," said Bill. "We were luckier there."

"It's not luck, Billy," I said.

"Then what is it?"

But Speaker was digging in, and Mathewson's first pitch to him gave me hope, a fadeaway that caught the outside corner at the high edge of the strike zone. He had a strike and he'd shown the breaking pitch, two important bits of success. Now he came in on Tris' hands with a fastball, and a late swing sent the ball straight up. It began to curve toward us, but I saw it would stay short of the rail, an easily played pop foul. Merkle was coming under it, and so was Mathewson. I heard the pitcher holler "Chief! Chief! Chief!" and saw Merkle obediently back away. But Meyers was only halfway down the line. The catcher sprinted and lunged for the ball, but it fell to earth amidst the three Giants, twenty feet from first, twenty feet foul. A jubilant explosion rocked the grandstand.

It had been Merkle's play; Mathewson had called him off. A million ball fans who knew nothing of Merkle but his burdensome nickname might have done the same, but this was Christy Mathewson in a World's Series game, and his mind should have held no portion of distrust for a teammate whose hit had given him the lead. I sensed that he realized the awful mistake, for he whispered a word to the first baseman before he snatched the ball from the grass, and Merkle nodded. But there was still the cost of the error to pay, for Speaker was alive at bat, and there was still but one man out.

Mathewson seemed resigned to the penalty; his fastball had no snap, and Speaker pulled it on a line into right field, where it bounded toward Devore. I was not surprised when the fielder made the wrong play, a hopeless throw home that allowed Yerkes to go to third and Speaker to second with the game tied. The Giants were infected.

"Bill, you're right," I said. "Let's go back to the family."

The Boston fans hooted when we left, and the two poachers, squatting in the aisle, scurried back into our seats. We trudged to right field while Lewis was purposely walked, and we reached the family in time to see Gardner hit a fly ball over Devore's head. Josh backpedaled to catch it and threw home with all his strength, but it was a forlorn effort; there was no chance to catch Yerkes, who tagged and raced down the line with the final run of the season.

Boston rejoiced, and we mourned. Bill, in tears, threw himself into his father's arms. I looked to the Giant dugout, and through the dance of the Red Sox I glimpsed Mathewson alone on the bench. Like Bill, Mathewson wept.

The youngster had the comfort of his father. Ruby hugged him and patted his head. "There's always next year," he said, and the boy, with his face pressed into his father's chest, nodded.

"Hey, Billy, let's have it," said Eli, his hand outstretched. "And two bits from you, sport."

October 4, 1919—at Chicago

										R	**H**	**E**
CINCINNATI	0	0	0	0	2	0	0	0	0	2	5	3
CHICAGO	0	0	0	0	0	0	0	0	0	0	3	0

BATTERIES: Cincinnati, Ring & Wingo;
Chicago, Cicotte & Schalk.
Winning pitcher: Ring. *Losing pitcher*: Cicotte.

THE WORLD'S CHAMPIONSHIP SERIES

	W	L	PCT.
Cincinnati (N.L.)	3	1	.750
Chicago (A.L.)	1	3	.250

Cincinnati leads best-of-nine series, three games to one.

FIVE

I LAST SAW Mathewson pitch on the first day of summer in 1916. He seemed a Gulliver in Lilliput as nameless little men swung at his pitches with insignificant result. They were collecting boasts for their old age: "I once hit against Matty."

There'd been another World's Series in 1913, the year after the famous Snodgrass muff; Mathewson had thrown a brilliant ten-inning shut-out for the Giants' only victory, then lost the fifth and final game with a crippled and erring team behind him. He was almost as good over the following season, but the club, those youths of '08, grew old that summer and finished ten games behind. The tumult that year had less to do with play on the field than the challenge of a new, third league. All the angry headlines of long ago were blazoned once again, the charges of desertion (by the owners) and enslavement (by the players), the astronomical leap in salaries, the declarations that traitorous athletes would never again darken a big-league dugout. Hal Chase jumped to the Federal League, raising his batting average nearly a hundred points and his bank account by thousands. Of all the established clubs only the Giants retained their veterans, but the 1915 season revealed that loyalty was a poor substitute for youth. The club that had won three consecutive pennants, the club of Mathewson and Marquard and Tesreau, of Meyers and Herzog and Doyle—and of Merkle and Snodgrass—lost eighty-three games and finished dead last.

That winter the near-bankrupt Federals pooled their resources and offered Mathewson twenty thousand dollars to pitch a single season for the franchise of his choice in their dying league. He refused. In '16 he was more often in the coacher's box than on the mound, and Giant pitchers had names like Perritt and Benton and Schupp and Sallee. As of old, the better club was on display in Brooklyn, in a sprightly and intimate new park of their own. A clutch of the family had settled in that borough of front stoops, living out my father's dream of thirty years before; the younger cousins adopted the local club as their own, and the rivalry on the field found a match in our holiday dinners and reunions.

Annually, Matthias and I went to the Polo Grounds on Opening Day, again in late June before our departure for Lake George, and once more in the final week of the season. I did not know that Mathewson's June assignment would be his last, but on the fifteenth anniversary of his St. Louis no-hitter I read of his retirement and advancement to Cincinnati to manage that forlorn franchise. The joke ran that professional baseball had begun in Cincinnati in 1869 and disappeared from that city soon after. A series of once great ballplayers had tried and failed to resuscitate the team. When Mathewson arrived he found but two young players of talent, Heinie Groh and Ed Roush, and one skylarking veteran returned from the Federals: Hal Chase.

The war in Europe killed the Federal League. Between the call for enlistments and the threatening draft there were insufficient athletes to stock the teams. Cities such as Buffalo and Newark aptly reverted to minor-league status, and most of the jumpers were forgiven their trespass. The '17 season was one of jingo excitement; war was declared in the first week of play, and ballparks staged patriotic extravaganzas almost daily, interrupting proceedings in mid-inning to sell war bonds. By summer no game would begin unless there was first a rendition of "The Star-Spangled Banner" by an operatic foreigner or a Ziegfield chorine of dubious voice. Zimmerman of the Giants, like Cincinnati's Groh, dropped the nickname "Heinie" and became Henry for the duration.

Our own community was not warlike; moreover, it seemed rank to ally ourselves with the Russian tsar and the French of the Dreyfus Affair against the Germans whose Jews were the freest in Europe and the Austrian Empire of that shining prince Franz Josef. Mister Sonnheim's firm had underwritten several German loans during the neutrality, but the Morgan bank held England's mortgage, and the Morgan bank made policy. Mister Sonnheim resigned on the day that Congress declared war. He became a frequent guest in our company box, often inviting old associates to take the sun there with him, and it charmed me to see these ancients, who'd weathered plunge and panic with equanimity, grow furious at an umpire's call.

The war put a brake on Arthur's projects but opened new and patriotic avenues. Our firm won contracts to coin military medals, and the agreement with the War Department provided automatic draft deferrals for our employees. The shop became crowded with male relations who had little to do and much to learn. Arthur organized training sessions in plant management and retail sales. He gathered blueprints for stores and factories and assembly units, drafted contracts for loans and bond issues, and collected lists of supply officers who would one day demobilize and need work. No allied government was as engorged by the war or as prepared for its end as Arthur's realm, Collegiate Jewelers, where he ruled as regent for our failing Uncle Sid.

Although the clubs raised millions for the war effort, although uniformed men were admitted without charge to any ballpark and enlistees were paraded around the field to stirring ovations, the schoolmaster in the White House thought the pastime frivolous in the deep hour of war. Already Wilson had closed the racetracks, and now he eyed the stadiums. To save the 1918 season the National Commission agreed to a foreshortened season and strong limitations on travel. With fewer games, the owners cut salaries commensurately—more than commensurately, the athletes claimed—and this, combined with the demise of the Federals, shrank salaries drastically. A solid professional infielder earned little more than we paid our apprentices. At the same time the sums passing hands in the grandstand multiplied many times over; with the tracks shut the ballparks offered the only alternative for an afternoon's adventure. More and more men of narrow faces and battered hats lined the ramps of the Polo Grounds, and in the box seats the high rollers took a new interest in the game. Eli welcomed them, for he could name his own odds.

The war years were ones of deep mourning for the family. My mother died in late summer of 1917, my father soon after, and in the winter my brother Sam succumbed to the tuberculosis he'd spent his medical career battling. Then in terrible succession came the influenza deaths of the summer of 1918: the gentle Mrs. Sonnheim, Cousin Saul's infant son, and finally our own five-year-old daughter Clara, the fruit of our European excursion of 1912. She began to show the symptoms aboard the very train we took north to escape the epidemic. From the station we took her by ambulance to the Lake George home and began a dreadful, sleepless watch. For two days the house heard no louder sound than the quick footsteps of the doctors on the stairway, and when on the third morning I heard the clamorous honking of an automobile horn I thought it was death arriving in raucous disguise.

It was Eli, and the car was new, a blaze of red paint and glittering steel spokes weaving up the long driveway in a cloud of orange dust. He saw me burst from the house waving frantically and honked the horn all the more gaily. I fairly threw myself into the roadster to muzzle the noise. His laughter died at my anguished report of Clara's condition, and after calling on Edith to express his concern he volunteered to keep company with Matthias. I sat with my daughter while Edith tried to sleep, and when she relieved me in mid-afternoon I walked to the pier and watched my brother and my son row in from their excursion on the lake. Matthias ran ahead while I took a slow walk through the woods with Eli.

"I'm glad you came," I said.

"Just a lucky chance," he said. "A month and more with Arthur is about my limit, especially with the war schedules for the trains. We had to share a compartment. I jumped ship in Albany, bought the car, and figured I'd motor up here."

"It's quite a machine."

"Twelve cylinders. I can get it up to seventy on a good straightaway." He reached up to snatch a pine cone off a tree and threw it to me. I took it one-handed and flipped it back.

"I saw Matty in Cincinnati," he said. "He looks well."

"He has them winning," I said.

"They'd be closer if he were pitching. I was surprised he took the job. I never figured him for a manager. I suppose he wants another title so he can go out a winner."

"He's done quite a job. When was the last time Cincinnati was in a pennant race?"

"Young Billy would know. I'm sure I don't. By the way, Matty asked if you'd do a ring for the club if they take the flag."

"It's in the contract with the Commission."

"But he asked if you'd design it personally. He has your drawing of the 'aught-four club ring on his office wall, likes to promise his fellows they'll have their own if they come out winners."

I caught sight of Matthias, a pantomime driver at the wheel of Eli's car. "It looks like you've been winning a bit yourself," I said.

"It's fish in a barrel these days. Since they closed the tracks there's a lot of dumb money afloat at the ballparks. They know as much about baseball as I do about . . ." He shrugged, unable to confess an equivalent ignorance. "It's going to be a whale of a stretch run, Mac's Giants against Matty's Red Stockings."

"And Boston. How are you betting?"

"I take it day to day," said Eli. As best as I or anyone in the family could judge, Eli had tamed his wagering to fit the limits of his income; he'd made no further calls on the firm's capital, and Arthur had his pledge that he'd never lay a bet with a client. On the western swings he saw to the clients' pleasures after Arthur dealt with business, and at home he'd become the company's official greeter for out-of-town buyers. Though his favorite haunt was still the ballpark, he knew the shows and the rooftop gardens where the entertainment began at midnight. He liked to flash the business cards imprinted with the strange title that Arthur had conferred upon him: Director of Public Relations.

No one wept for Clara more bitterly than Eli, and no service was more mournful. Our grief for the older generation had been tempered by the knowledge that deep age is death's season; my brother Sam's passing had about it an air of heroic sacrifice, and the infant buried the month before was a life spirit without personality. Clara's was the cruelest death, and I was benumbed by it, hardly able to remember the sorry train ride back to the city and the doleful funeral in the terrible August heat. For the week that followed I did little more than hold Edith's hand and weep with her. On the eighth day we felt compelled to visit the gravesite, a green hillside in rural Queens which was marked for us all. On the return ride Edith broached a new matter: her father had invited us to share his large and now empty house in Turtle Bay, and Edith thought well of the idea, for our home seemed too full of Clara. She wondered if I could work in strange surroundings and I assured her that I could, remembering the heat of an Illinois night and the ruby I'd drawn as its symbol.

At home, Matthias had a note that an office runner had delivered for me that morning. I opened it and read:

> Jack—
>
> I'd much prefer not to trouble you at this time, but Eli begs you to attend a meeting at the shop this afternoon at one o'clock. Carl will be present, but with Uncle Sid not in health and unable to be there Eli hopes you will join us.
>
> Arthur

It was after one, but I welcomed the opportunity to leave the set of mourning that our house had become. Twenty minutes later I was at the shop, entering beneath the banner that proclaimed it "The House of Heroes." I took the lift to the topmost floor and walked to Arthur's office, once my own. I barely acknowledged the sympathetic expressions of the staff. As I reached for the door it opened, and I was face to face with Eli.

"Thank God you're here, sport!" he said, throwing his arms around me. I looked over his shoulder and saw Arthur at his desk, dressed in the brown uniform of his captaincy with the ribbon connoting distinguished service to the supply corps; the rank was honorary, but the profits were not. Arthur's face was all anger and contempt. I caught Carl's eye as he stuffed papers into his briefcase.

"Come in, Jack," said Arthur. "I'm sorry you have to be part of this."

"Part of what?" Eli released me, and I straightened my jacket. Carl handed me a paper which bore the letterhead of the National Commission of Baseball. It read:

August 8, 1918

Eli Kapp
Collegiate Jewelers, Inc.
205 Grand Street, City

My dear Mr. Kapp:

Kindly be present at these offices on the twelfth of August at ten o'clock in the morning to respond to questions in the matter of player Chase of the Cincinnati ball club.

Very truly yours,
John Heydler
President (Acting)
National League

"They've been intercepting my mail," Eli said.

"I had a call from Heydler yesterday," said Arthur, dismissing the complaint. "He told me what the matter was. This is deep trouble, Jack. It could foul our entire relationship with the National Commission."

"I'm at a loss," I said. "What is this 'matter of player Chase'?"

Carl took the letter from me. "Player Chase," he said, "has apparently bet heavily on clubs opposing his own, and he's won those bets far more often than not."

"And guess who's been placing the bets for him since we left Cincinnati on the swing?" said Arthur. "And guess who used a check from Hal Chase to purchase an automobile in Albany last month?"

I moaned. Eli stood with his head bowed.

"Chase will be placed on suspension today," said Arthur. "It would have happened long ago but for two factors. First, there was no hard evidence. Chase is discreet. More so than his agent, at any rate. The second reason is that Mathewson has refused to suspend him. He is either the most credulous man in America or its greatest model of Christian forgiveness. Perhaps both. However, the Cincinnati ownership is not so trusting. They put a detective on the case, who overheard certain discussions between

Chase and our brother. Eli was followed from the time we left Cincinnati. He had a check from Chase, stakes money, but Eli's credit still holds in such centers as St. Louis and Chicago so he had no need to cash the check. He held it until Albany, where he endorsed it over to the Lauder Automotive Agency in partial payment for a twelve-cylinder Morgan roadster. He paid the rest in cash. Then he purchased a Western Union money order with additional cash and wired it to Chase. Do I have all this right, Eli?"

Eli nodded.

"Jack, this is all quite secret," Arthur continued. "There'll be no charges, no legal investigation. They do want to get Eli on the record as part of the case against Chase, so they've invited him to testify on Monday."

"Not testify," said Carl. "It's not an official proceeding. The phrase is 'respond to questions.' He'll do that, freely and openly."

"What will he say?"

"That he's never acted as Chase's agent. That he has a friendship with Chase of ten years' standing. That Chase's check to him was in repayment of a loan made in 1910 which Chase used to purchase an automobile, so that it was merely a neat turn for Eli to use the repayment to buy another. As for the monies he wired to Chase, Eli will state that it represented fees for various promotional services Chase has rendered to the firm, which fees were held up while Chase was in the Federal League."

"Officially we considered his jump to the other league an action detrimental to baseball and to our company," said Arthur. "Heydler will like that."

"Eli will further state that he had Arthur's approval to make the payment to Chase, seeing that the relationship had been restored during their visit to Cincinnati. We'll produce evidence that the cash was drawn from our Albany store. Thus Eli's response, for the record."

"Of course I've told Heydler the facts of the case," said Arthur. "All this is window dressing for the official report. I'm going to have to scramble to save our relationship with the Commission, but it will be worth it. Jack, I'm putting together a tremendous deal.

There'll be no more payments to individual ballplayers for endorsements, just a lump sum to the Commission in return for free use of any player's name. It'll save us thousands, actually tens of thousands, after the war."

"Arthur, I really don't think that Jackie's interested in the business end of things right now," said Carl.

Arthur winced. "I'm sorry, Jack. That's the very reason I didn't want you present in the first place. It was Eli's insistence."

I turned to Eli. "Why?" I said.

He looked at the floor. "I'm afraid, sport. I need you here to hold me up."

"No, no, no. Why did you do this thing?"

"Because it was a sure thing, sport! Only a fool would pass on a sure thing. And I didn't take any money from the family! I thought I could do whatever I wanted, as long as I didn't take money from the family! Wasn't that the idea?"

"Didn't take money from the family!" Arthur shouted. "You put hundreds of thousands of dollars in jeopardy! We're tied to the National Commission by contract! What hurts them hurts us! How can you be so oblique that you don't see that?"

"They're making me lie, Jackie," said Eli. "I've never told a lie in my life, and now they want me to go to the Commission on Monday and lie up and down for the record. I don't want to lie. I don't want to go to jail for lying."

"What a remarkable moral stance," said Arthur.

"Nobody's going to jail," said Carl. "I repeat that Monday's examination is not a legal proceeding. Perjury charges cannot obtain from this inquiry, even if Eli swore he's been fighting in France the past year and has never been to Cincinnati in his life. I'd never instruct you to lie under oath in an official proceeding, Eli, but if Chase challenges the suspension there will be endless official proceedings, and we have to dance this dance to avoid that."

"Heydler knows the truth of the matter," said Arthur. "He has us over a barrel, and it's going to cost us some percentages in future dealings. We have to follow his script or lose the Commission's business completely."

"And what about Mathewson?" I said. "How could you do this to Mathewson?"

"Do what to Mathewson?" Eli and Arthur said together, and Arthur continued, "We haven't done business with Mathewson for nearly six years."

I ignored Arthur to question Eli. "One of Mathewson's ballplayers proposed to you that he would purposely throw ball games, and not for a moment did you think to tell Mathewson about it. You told me at Lake George that you'd talked with Mathewson. How could you look him in the eye, knowing what you did?"

"Matty's a swell fellow," said Eli, "but Hal Chase is my friend. I couldn't rat on a friend."

Arthur hit his desk. "That's it, Eli. You're done, you're finished. I don't want you in this building ever again."

"Just a moment!" I said. "I should think that Uncle Sid would have the final word——"

"Uncle Sid is dying. He'll live the summer and die in the fall. If you wish to burden him with this contemptible matter in his illness I can't stop you, but I swear that as of the day he dies these doors are shut to Eli."

"Eli will still have his stock and the income it yields," said Carl. "He'll never starve. I'll see to that."

"He can join the Army for all I care," said Arthur. "He can go and get his head blown off making the world safe for democracy. It's the least he can do after fouling things up as he has."

Eli, of course, did not join the Army in contrition for his sins. That was Mathewson's way, and with two volumes of evidence before him, one an official record, the other a private report from Heydler, he issued a statement to the press.

"Undesirable elements have apparently conspired to influence the play of certain members of my ball club and thus the outcome of our games. I do not know that they have been successful in their schemes, but a manager has a responsibility to protect his players from such contacts, and evidently I have failed in that regard. The players and ownership, and especially the fans of the Cincinnati ball club, have my deepest apology and most sincere regrets.

"Today, all Americans are engaged in a course of heroic purpose.

Had I not undertaken certain contractual obligations before war threatened I would have joined the cause long before this. Having resigned today as manager of the Cincinnati ball club, I am now free to enlist in our country's service. God bless our country, and God bless you all."

I was in black when the Armistice rang. The family had gathered to mourn Uncle Sid in the temple he'd helped to build, while outside, in a ghoulish counterpoint, his wife Riva's coffin was disinterred to ride with his to that distant burial ground which poor Riva had never seen. Our patriarch died as his world ended in surrenders and abdications, and he was buried on the day, at very nearly the hour, that the bells announced the new world's dawning.

My silent prayer at the gravesite was that the deaths might now be done; I spoke the thought aloud in my toast at the New Year of 1919, which we celebrated in my father-in-law's echoing mansion, now our own home. As week after week the returning armies paraded up the cold winter Avenue I thought much on the irony that had come to mind at all our family funerals: I knew no one who'd been killed or even wounded in the Great War, yet the time had been shrouded in mourning. Then, in February, I read the news about Mathewson.

I came upon it accidentally, a small item in the corner of a page given over to our advertisement for St. Valentine's Day selections. "MATTY" IN BELGIAN HOSPITAL, the headline read, and then a single sentence:

> A report has reached Allied headquarters in Paris that "Christy" Mathewson, late of the New York Giants baseball club and former manager of the Cincinnati nine, is in hospital in Chimay, in the southern salient of Belgium.

That was all. I hurried to the corner stand and bought all the city's newspapers, fumbling to scan one after another while carrying the rest. One printed the identical story without a headline, and another repeated it with an additional paragraph:

According to an AEF spokesman, Mathewson commanded a weapons disposal unit in that sector. The hospital is a field unit that treats victims of poison gas, the spokesman said.

There was no news the next day, nor the day after that. Were we so far from the time when the *World* would publish an extra to announce that Mathewson had risen from his sickbed? I called the Giants offices and identified myself as the designer of the club's championship rings, but a secretary told me she had no information on the matter. She pledged to remember me if any word was received, but she did not, for the next I read of the pitcher it was spring, and he was recuperating at his Pennsylvania home. A photograph ran with the story, and I could not believe that the man pictured was Mathewson, so old did he appear, his face so deeply lined, his hair so plastered with gum, and his eyes so dumbly staring beyond the camera into an immense and vacant sky. I asked Arthur to inquire of the War Department if Mathewson had earned a decoration, but he learned nothing of the sort.

"If anyone knows what happened they're not telling," Arthur informed me, "and when I asked about a decoration the man was startled at the thought. No one wants to talk about it, Jack, and as for Mathewson himself . . . I don't know that he'd appreciate anything from Eli's family."

Certainly I'd thought of that; it was the reason I'd not written him directly. I'd hoped to approach him in the old way, with a precious tribute. Now I attempted a letter, but found myself writing more about my own sorrows than his; I discarded it, tried again, failed again. I was at the door of the parlor car, watching him through the glass, incapable of speech, and now I had no recourse to stylus and ink. I could pray that time would sort the matter out, but when I looked at the photograph I wondered how much time was left to him.

And, as if kicking dirt onto Mathewson's yawning grave, McGraw signed Hal Chase to play first base for the Giants.

I did not attend Opening Day at the Polo Grounds. Carl gratefully accepted my ticket and afterwards asked why I'd given it up; I answered that it had to do with "the matter of player Chase." Carl shook his head.

"You have it wrong," he said. "Chase contested the suspension. He sued the Cincinnati club for his blocked salary. Something had to be done for him, or else the whole story would become public. McGraw finally volunteered to hire him."

"Only McGraw would trust such a man," I said.

"Trust doesn't enter into it," said Carl. "Mathewson may have thought to shame Chase by enlisting, but a man like Chase can't be shamed. McGraw exercises a different sort of authority. He told Chase that at the first sign of any funny business he'd brain him with a baseball bat."

Despite that canker on the Giant roster, Mathewson returned to the Polo Grounds in June. Perhaps he needed to hear the cheers, or earn the money, or forgive Hal Chase in a personal interview; for whatever reason he donned the old uniform, and inning by inning he walked the way from the coacher's box to the dugout with slow, measured determination. He remained on display through an entire homestand, but when the club took to the road—as did Arthur, without Eli but with half a dozen trainees—Mathewson went north to Saranac Lake for further treatment in the sanatorium there. The place was not eighty miles from our Lake George retreat, and I had a slight acquaintance with its director, a legacy from my departed brother Sam. I wrote to the man, invoking Sam's memory and claiming an old business relationship with his patient; I asked if I might visit the pitcher. I received a cursory response. Mathewson's physicians thought a visitor inappropriate. I couldn't guess if Mathewson had been told of my request. I didn't write again.

My autumn call at the shop had developed aspects of ceremony. At ten there was an assembly of apprentices and a flowery introduction from Arthur; I spoke on "The Elements of Design and Construction for the General Public." Then I toured the shop with Ruby, learning the capabilities of the newest machinery, and lunched with the sales force to hear their estimation of the market. In the afternoon a formal meeting of the board was convened.

This was the first since Uncle Sid's death, and I said a silent kaddish for him as Carl read the bylaws and explained the structure of the board. Eli, who'd been willed equal voting rights, was not present; Arthur was installed as chairman without opposition.

Arthur pushed for action. Using huge multicolored charts that mapped authority and responsibility, giant blueprints of stores and factories, accounting sheets dotted with spectacular sums, he put forward his whole plan for national expansion. He explained the initial steps he'd taken on his own authority and ended by comparing the state of the firm to that of the Allies on the eve of their final push against the foe. Then he moved adoption of the entire plan and sat down at the head of the table.

We voted for delay. Uncle Sid's sons, Ruby and Saul, were frankly overwhelmed by Arthur's grand design, and Carl voted Eli's proxies against the plan. "I'm instructed to vote against any motion of which Arthur is the author," he said. I abstained, as I'd vowed to do on any ballot that wasn't unanimous, while Arthur and Carl voted together.

"The motion fails of a majority and is therefore defeated," Carl announced, and the matter was put off for further study.

Ruby and Saul were probably guilty of every offense Arthur then charged—smallness of vision, failings of courage, insufficiency of gratitude—but they were also entitled to caution as they tested the burden of their new responsibilities. "Give us some time to look at all this," said Ruby, eyeing that mountain of charts and plans. "I'll feel better voting on it a year from now. I move we adjourn."

The motion was passed, and Arthur grabbed my arm and pulled me to the corner office he'd claimed as his special inheritance. His war medal and citation hung on the wall, opposite a drawing of the Boston store and its descendants in Albany, Hartford, Providence and Cincinnati.

"How could you let me down like that?" he shouted after slamming the door behind him.

"I told you what I was going to do," I said. "I've no intention of getting in the middle of all this. Besides, even if I'd voted with you the motion would still have failed on a tie vote."

He sat at his desk and swiveled his chair to gaze out the window. "How much do you want for your stock?" he asked over his shoulder.

"Why, it's not for sale."

"Why not? You don't intend to do anything with it."

"Aside from supporting my family and sending my children through—my son through school, no, I don't have any plans for it."

"You could exchange your shares for preferred stock."

"What's that?"

"Non-voting shares, the sort that most of the family has. You'd receive higher dividends."

"I'll consider it, but I'm inclined against it," I said. "If Uncle Sid had thought that a better scheme, he'd have——"

"Damn it, Jack!" He turned to face me. "There's a timetable for all this, don't you realize that?"

"Perhaps you've exceeded your authority."

"I've done nothing of the sort. It isn't authority I've taken into my hands, it's responsibility. I've kept this firm alive through Uncle Sid's dotage and Eli's machinations. I've seen your father-in-law for advice on the financing, did you know that? It's his timetable, not mine. It has to be done this way."

"You'll have all the authority you need after Ruby and Saul study the plan."

"I don't have time for that! I'll lose options on store locations, I'll have to cancel orders for machinery, and half the people I want to hire will find other work! Please, Jack, give me your vote on this. I'll give you anything you want."

"I have everything I want. Everything that you could give me, at any rate."

"You're so blind! You don't see what's out there waiting to be grabbed. Jack, what you have now is small cheese compared to what's possible. Yes, you're a rich man, so are we all. The next generation will be quite comfortable as well. But after that! Will Matthias' children be rich? Did Uncle Sid build all this so that the family could live well for two generations and then slowly sink back from whence we came? Do you want your grandchildren to disappear into the crowd, your great-grandchildren forced to start all over again?"

"It might be better for them."

"But the opportunity is now! 'Once in a lifetime' doesn't begin to describe the moment. Europe's finished, bled white. It'll take a century to revive, if ever it can. You've been there, you've walked through the ruins, you've seen how a civilization can stagnate for centuries. This is America's hour. If we stay out of dying Europe's wars and build our own empire it can last a thousand years! There are dukedoms to be won, family fortunes to last an eternity! We're that close to it, and what stands in our way? A pair of gutless cousins, a maniac who's been out to cut me down since I joined the firm—and you, the Great Abstainer." He shook his head. "Perhaps the family's unworthy. Perhaps that's the lesson. There are other families, you know. Some of them have attractive daughters. I'm thirty years old, time to marry."

"You'd resign?"

"I might, if an offer came."

"And then what befalls this family? What of its grandchildren then?"

"That is entirely up to you. I can reconvene the board if you'll vote the right way."

"It would still be a tie vote."

"I can initiate certain actions on my own. The vote would then be on a motion to rescind them, and it would fail—on a tie vote."

"But you can't reconvene the board just like that."

"Can't I?"

"You cannot. I gave half an ear to Carl's reading of the bylaws. You can't convene an unscheduled meeting of the board unless a clear majority seeks it, or unless a board member dies."

Arthur unsnapped his silver case and lit a cigarette. "Then I'll have to kill someone, won't I?" he said, laughing.

"Arthur!"

Seeing my expression and remembering the death I carried with me he become abject with apology. He walked me to the lift with his arm about my shoulder, assuring me that he respected my stance on the vote but urging me to consider exchanging my voting shares for the other sort. As the doors to the lift closed he smiled at me, but as I descended I heard his fist beating a tattoo on the wall above me.

The following day Edith had a note from him.

September 23, 1919

Dearest Edith,
Might you permit your menfolk to visit Cincinnati for the first two games of the World's Series? We've arranged a supper with the National Commission on the eve of the Series, and Matthias should find that tremendously exciting, an opportunity to shake hands with all his heroes. Moreover, I don't know that he's ever met his Cincinnati cousins. Ruby's boy Bill will be along to keep him company. If it doesn't overly interfere with his studies, I'll be happy to make the arrangements. They'd depart on the 29th to meet me in Cincinnati, and return on October 3.
I thought that Jack's talk to the apprentices was outstanding, as always. You really did marry the pride of our pack.

As ever,
Arthur

I hadn't planned to go to Cincinnati; neither the formal supper where the championship rings would be unveiled nor the Series itself appealed to me. I'd seen to the manufacture of a Cincinnati club ring for Mathewson, for the team was his creation and its new manager, Pat Moran, but a Joshua to Mathewson's Moses; that done, I had enough of it. Nor did Edith have enthusiasm for the idea, and we kept it from Matthias, not to disappoint him, but on the following afternoon his cousin Bill came to call, all excitement at the prospect. When we sent Matthias to bed that night we promised we'd discuss it over supper. We did so, drawing Edith's father into the matter. Mister Sonnheim thought well of it: travel, excitement, a break in the schoolboy's routine.

"Perhaps you'd prefer to take him," I said, knowing his delight in shepherding Matthias to the Polo Grounds.

"Oh, no, it would be rather too much of a task," he said. "Ten years ago surely, perhaps even five, but not today. You hardly travel at all these days, Jacob. Why not see if your old haunts remain?"

"I prefer the memory," I said. "Everything was simpler then."

"Illusion," said the old man. "It's our capacity to see complexity that increases, not complexity itself. We have a name for that capacity. We call it 'wisdom.' "

Edith laughed, and joked that her father aspired to the reputation of a sage.

"I'd prefer the reality to the reputation," he answered, "but I make no claim to either. What sort of contest will it be, Jacob? Are the teams equally matched?"

"Chicago is the more experienced club," I said. "You saw them yourself two years ago when they beat the Giants in the Series. Cincinnati's younger, and not as deep in talent. I think their chances would be better if they were still in Mathewson's charge."

"Yes, poor man. He looked ancient, there on the coaching lines in June. And he was such a vigorous lad. I only met him the once, you know. At the club. What an excitable chap he was! I had to calm him, twice. There were heads turning, that sort of thing. Surprising in a boy of his family. There was a bit of the ruffian in him, don't you agree? A bit of his manager, McGraw. It was in his eyes. A long time ago, of course." He smiled at Edith. "Perhaps today he aspires to the reputation of a sage."

How remarkable the old man's memory of that supper! I'd thought Mathewson the perfect gentleman, but Mister Sonnheim was left with a vision of impetuous energy. Strange, I thought, and then Edith was asking her father if he really thought the Cincinnati excursion was for the best.

"I do," he said, "and not alone for the occasion. He has relations in Cincinnati, and one oughtn't pass up the opportunity to meet more family. Yes, let him go, by all means."

We boarded the overnight train at Pennsylvania Station late the following Monday. Bill, a man-child at seventeen, played parent to my ten-year-old, but he was not so far from the enthusiasm of his childhood, for he still kept books of statistics, and these were their study on the train. National Leaguers both, they searched for signs of Cincinnati's advantage. One might argue that their pitching was deeper than Chicago's, but the Reds had no single master to match Cicotte of the White Sox. Little Eddie was a control artist who'd blossomed in the autumn of his career. He'd won twenty-eight games in 1917 and beaten the Giants in October, and though he pitched unsteadily in '18 he returned to splendid form in the season just ended, winning twenty-nine times. No one else in baseball was near him. In hitting, the clubs hardly compared: Cincinnati's Groh and Roush were worthy, but Chicago! Joe Jackson, Eddie Collins, Buck Weaver, Ray Schalk, Nemo Liebold—nightmare names to American League pitchers, and each a defensive standout as well, like their teammates Gandil and Risberg and Happy Felsch. Cincinnati was a solid ballclub, but in the Series they simply didn't figure. The boys were gloomy. I reminded them that whatever the tale told by print on paper the players were flesh and blood, and starting out even.

Our store in Cincinnati was manned by the cadet branch of the family, and above it Arthur kept an apartment. This was fortunate, for the Series overburdened the city, and hotels as far north as Dayton were booked full. Not since '16 had the sporting crowd traveled free of wartime restrictions, and the Series of 1919 had aspects of an ultimate celebration of the peace. Appreciating this, the National Commission had extended the competition to nine games. Only the very first Series had offered such a course, and the town was alive to it: red pennants fluttered from every lamp-post, and windows were bedecked with banners of encouragement. It was the first post-season contest to excite the city in the half-century since that first club had determined there was a living to be made playing a boy's game. The peace, the extended format, the fiftieth anniversary of professional ball with the championship at stake in its birthplace—it seemed beyond coincidence and hinted that forces were at work to mark this World's Series above any played before.

Our Cincinnati cousins, those strangers of our own blood, greeted us at the store and opened the apartment to us. They were cheered to find that Matthias and Bill harbored no sympathy for the American League. Matthias protested that he'd never even seen an American League game; the Yankees, he insisted, were mere tenants at the Polo Grounds, playing at McGraw's sufferance amid the echoes of the Giant crowds. We were told that Arthur awaited us at the Hotel Sindon, where the National Commission was headquartered. We hastened there and found Arthur a picture of assured command. He stood in the center of the grand ballroom issuing orders to the staff, approving a floral bouquet, instructing the busboys in the art of folding napkins. He had a pass personally signed by league president Heydler to permit the boys to watch the home club work out at Redland Field, and they dashed off to the ballpark. I knew my task, and stripped to my shirtsleeves to polish the rings that would be the centerpiece on the dais that evening. At intervals word would pass that yet another famous personage was checking in at the desk, and the staff would drop whatever work was at hand and rush to the terrace to watch the event.

At noon came a shout that the White Sox were arriving. I joined the crowd on the terrace encircling the lobby. There they were! Such a collection of citified dandies, such a display of white cuff extending two inches below the sleeve, such an array of pearl stickpins and diamond rings! Our trophies would merely gild these glorious lilies. They moved with grace and restless energy as they jousted and joked with one another; they moved with confidence and delight in themselves and the occasion; they moved as champions, who shared a sacred mystery denied to those of us who stared at them from our high remove. We would witness the event; they would shape it to their will.

When the boys returned we left Arthur and motored to a farmhouse ten miles out of town for dinner with the cousins, then returned to the apartment to change to tuxedos and hear Matthias' plaint that he was too old for short pants. A chauffered limousine took us back to the Sindon. The ballroom gleamed with crystal, and the rings glittered in their frame. We watched the grand procession of dukes and earls and sheriffs enter in irregular groupings. Garry Herrmann, doubly proud in his double capacity as chairman of the National Commission and president of the Cincinnati club; Ban Johnson, bulky and aging but confident that his league would preserve its decade-long dominance in post-season play; John Heydler, his politician's smile hiding a politician's secret—these were the dukes, as the club owners were the earls, each host at his own table: Comiskey of the White Sox, his profile cut from an antique Roman coin; Frazee of Boston, with George M. Cohan as his guest—he'd made a fortune backing Cohan's shows only to lose it, and much more, on a dozen others; the new Yankee owners, "Captain" Huston and "Colonel" Ruppert, their ranks as titular as Arthur's own, though how a brewer like Ruppert might earn the honor was a surpassing puzzlement. These and a dozen more ranged about the ballroom, eyeing one another with envy or contempt or tolerant amusement. And the managers, the sheriffs obedient to their liege lords' commands: Chicago's Gleason and Cincinnati's Moran, the week's combatants; Jennings, Griffith, and gaunt Connie Mack, who shared a century of experience and a security beyond the dreams of their younger colleagues; the new boys, Huggins of New York and Bezdek of Pittsburgh and husky Branch Rickey of St. Louis, some attending their first Series as managers, and some, if only they knew it, their last.

And McGraw. He entered with his erstwhile teammate and current rival, Wilbert Robinson, on his left arm and on his right a boy, a beautiful boy, a Mercury, a Mathewson drawn to smaller scale. The youth showed a shy reserve in this company of familiars and seemed unaware that for the moment he was the object of every glance. I heard him called McGraw's new boy in a roster that began with Mathewson, and I picked his name out of the murmur in the room: Ross Youngs, the Giant right fielder. He seemed a yearling fawn, with a fading memory of awkward youth and a proud precognition of mature strength and vigor. He walked at McGraw's pace, a quarter-step behind him, and when the manager stopped to greet an old friend or foe the boy took a backward step, bowed his head, and clasped his hands behind his back. I felt a desire to draw him in that pose, perhaps looking into a crystalline pool, intrigued and confounded by his own aspect. They sat at the Giant table, where McGraw held court while many found opportunity or excuse to pass within range and study Youngs. He was the only player in the room; the festivities were barred to active athletes and to the press, but McGraw made his own rules. I watched Arthur take the boys in tow and introduce them to Wilbert Robinson, and I saw shy, frightened smiles gloss their faces when Robinson turned and presented them to McGraw at the adjacent table. The boys returned to their places dazed and glowing. No greater excitement could follow that; the food was mediocre, the speeches boring. Neither protested when I checked the hour and shooed them to the limousine and the apartment over the store.

I stayed. I'd begun to count the championship rings on the hands of those present, and when the figure reached three dozen, and when my glass was empty for the fourth time, I thought of the indirect but certain route I'd traveled to this place and time. Would I be here if my parents had yielded twenty years before and allowed me to pack off to Altoona? Fewer than half the company had ever played the game for money; the owners and their lieutenants were for the most part promoters or businessmen, and their guests were men of enterprise who carried no hint of athletic form. The rest were survivors of the dreadful attrition the game enforced upon its own. Of the millions of boys who ever put bat to ball, how many had signed a professional contract in the fifty years this occasion celebrated? Twenty thousand? Of those, how many had advanced even a single stage, let alone to the big leagues, and of each year's rookie crop how many played a second year, or a third? Fewer still found a place in the game when their playing days were done, fewer than two hundred; that was the survival rate out of millions. Sad, too, that the longer the career the more desperate the condition when it ended, for what else did such men know but baseball? How long could their fragile celebrity support them once outside the game? Precious was a manager's posting in a rude backwater of Class D ball, for without that the forgotten men became shadows, walking ghosts, figures to trigger a memory of days past but irrelevant to the present. They ended as statistics in Bill's neatly filed books, mere measurements against today and tomorrow, gravestones. For twenty years I'd grieved for what I'd lost when I gave up the dream of a professional career, but what might I have today if I'd followed it? Not Edith, not Matthias, not my home and my work; not this chair in this emptying ballroom, nor the fine effervescence I poured into my glass, which I raised to the display of rings on the dais.

The bottle was empty. I stood up and walked unsteadily around the perimeter of the ballroom, searching for a table with a half-filled magnum. Here and there clusters of men lingered to enjoy cigars and conversation. I saw the Yankee owners and their manager, little Miller Huggins, in close conference with Boston's Harry Frazee and his field boss, Ed Barrow. They cut off their talk as I passed and spoke more quietly as I drifted toward McGraw's claque.

"Look at that," McGraw was saying as he pointed at Ban Johnson, slowly shuffling away with the support of two younger men. "The room's beginning to smell better already. He still runs a minor league. A strong one, I'll give them that, but a minor league nevertheless. Chicago's their only first-rate club, now that Connie's unloaded his fellows."

"Boston's not bad," said Wilbert Robinson, who'd moved to the Giant table. The Brooklyn space was empty, and a bottle there not yet drained; I poured a glass and sat with the back of my chair tipped against the shadowed wall. I had McGraw's profile in focus, and a three quarter view of Ross Youngs, who sat beside the manager and watched him, expressionless.

"Boston was good enough when they beat us," said McGraw, "but, hell, Robbie, when you played them Speaker was gone, Yerkes, Stahl, Cady. Joe Wood. We had to face Smokey Joe."

"And we had Shore and Mays and Ruth against us," said Robinson, recalling the Series of 1916. "I'd match Ruth against Wood any day. The kid would have shut us out if his center-fielder hadn't given us a run in the first, and he threw blanks for the next thirteen innings."

"If he's such a pitcher," said a man as he lit McGraw's cigar, "why does Barrow play him in the outfield?"

"Home runs," said Robinson. "Twenty-nine this year. Twenty-nine!"

"I know the number," the companion said, "but it doesn't mean much at Fenway. You can stand at home plate and spit over the wall in that park."

"Ruth bats left," McGraw said.

Someone whistled. "Twenty-nine home runs, batting left, in that park?"

"That's right," said McGraw. "What do you think he could do at the Polo Grounds?"

"Why? Are you after him, Mac?"

McGraw snorted and pointed across the room. "See Frazee and the Yankee brass? There's a deal in the works, that's why the managers are there."

"I hear Barrow's coming to New York," said Robinson.

"Huggins is out? Damn shame, he's a good man. National Leaguer."

"Hug will stay as Yankee manager," said Robinson. "Barrow's coming in over him."

"Over him! What kind of deal is that?"

"Hug will supervise the club on the field, and Barrow will get him the players," said Robinson. "It's Ruppert's idea. Ruppert realizes he doesn't know beans about baseball. He's calling Barrow his general manager."

"'General manager'!" McGraw spat. "Ridiculous. A manager's a manager and an owner's an owner, unless it's all the same man like old Connie or Clark Griffith. The best operation is a one-man show. You know that, Robbie."

"That's why I turned it down when Ruppert gave me a nibble," said Robinson. "Nice thing about it, though. It gave me some leverage with Brooklyn. They're giving me a piece of the action."

"Congratulations," said McGraw. "How'd you like to invest in a bar with me?"

Robinson laughed. "Those were the days, Mac. The old place in Baltimore is still standing, you know. Someone mentioned it to me today."

"Jack Dunn? I saw him, too. He let me in on a little something about Babe Ruth, by the way." McGraw winked.

"What's that?"

"Nigger blood. His mom had the taint, half of Baltimore knows it. You can see it in that moon face and those liver lips."

"And you'd let him play in your park?" said one of the company.

McGraw shrugged. "I've snuck in a few in my time. Called 'em Cubans or Indians. They can play ball."

"Is that the deal, then?" said another man. "Barrow's coming to New York and bringing Ruth with him?"

"More than Ruth, I'd guess," said McGraw. "Harry needs the money up there in Boston. He's got dollies stacked in apartments all over the league. The overhead is killing him."

"He'll draw a crowd," said Robinson.

"Ruth? Yes, and I'll get a cut of it. Maybe I'll boost their rent. It'll last a season, maybe two, and then the pitchers will wise up or Ruth will drink himself out of the league, like Donlin or Bugs Raymond. He's a boozer, a skirtchaser too. You don't build a team on his likes."

"Still, twenty-nine home runs in Fenway . . ."

"And three strikeouts for every home run," said McGraw. "That's no way to win. Hell, the Giants were doing that when I came to New York. Couldn't bunt, couldn't run. They just swung from the heels and prayed they'd connect."

"That big guy can win a game a week with his home runs, though."

"All right, a game a week," said McGraw, "but look at Ross here. He's worth three runs a game, the one he'll get you at bat and the two he'll save you in the field. Ruth isn't half the ballplayer that Ross is. This boy can do it all, run, throw and catch, bunt, spray it to all fields. Look at him! Stand up, Ross. Look at that body! Perfect for a ballplayer. Ruth isn't built for the game, with that big belly and those spindleshanks. No speed, no glove. Ross is a ballplayer! Look at that chest and those arms. Take off your shirt, Ross."

Silent, obedient, the boy doffed his jacket and undid his shirt studs. He removed the shirt and the undergarment in a swift, graceful motion and stood barechested at McGraw's side.

"Now, does that look like a fat half-nigger to you?" said

McGraw. "And the legs, wait 'til you see the legs. Take off your trousers, Ross. The skivvies, too. Now get up here on the table."

Naked, the boy sprang onto the table and made it his pedestal, standing in repose, his body white, his face still as a statue's. The men at the table leaned back in their chairs, studying him as if he were a slave at auction.

"Look at that body," said McGraw in a low voice. "If I'd had a body like that I'd have been twice the ballplayer I was. Look at him!"

The boy turned slowly; the men gazed upon him. At other tables people rose and walked quickly toward the display, gathering in a growing circle. Someone began to applaud and was joined by others. McGraw mounted the table.

"Gentlemen, this is Ross Youngs!" he announced. "Ross Youngs, right fielder for the New York Giants. This is my champion, my next champion! And you can take your bloated Babe and open a sideshow at the Polo Grounds," he shouted through cupped hands at the table across the room, "while this boy is bringing home a title!"

There were cheers, and McGraw took the boy's wrist in his hand and raised his arm in a champion's pose. He clapped him on the back and followed him off the table, moving his round body with a faintly remembered grace. Youngs slipped into his clothing amid a chatter of old men's voices. I watched him, and then I watched McGraw watching him; my vision blurred, and when I closed my eyes I saw a photograph of a dying man's face. I looked up and saw McGraw with his arm around the boy's shoulder, and I growled and lurched forward, the bottle in my hand. I raised it like a club, the champagne running down my sleeve, and I took a step toward McGraw, but then Youngs was in front of me, his hand was on my chest, and I was falling backwards onto the floor and the bottle was skittering toward the wall. When I opened my eyes again I saw Arthur's face looming above me.

"Jack, Jack, take it easy now, just stay here. Waiter, bring me some black coffee right away, and, Jesus, a towel! You're bleeding, Jack. Just relax, I'll get you up to my room, we'll get a doctor . . ."

*

When I woke up the next morning my first thought was of the boys, but a note I found on the bedtable reassured me: Arthur had spent the night at the apartment while I'd slept in his bed at the Sindon. I was to meet all three for lunch and then the ball-game. I would have preferred to stay in bed—the more I thought of the previous evening the more I felt I ought never leave the room—but of course it was impossible, for I had the obligations of a father and an escort. Their enthusiasm failed to rouse me to the excitement of the game. It was a sloppy affair from the outset. Cicotte's first pitch hit the Cincinnati batter, Morrie Rath, square in the back. The teams traded early runs, but in the fourth inning Cicotte collapsed in a confusion of extra-base hits, errant throws and failures of skill and omission. Cicotte left the scene to a mocking and grateful ovation from the home crowd, and at the last out, an hour later, the fans surged onto the field to praise their favorites with hallelujahs.

Arthur had appointments after the game; we three waited an eternity at an overpacked restaurant before a harassed headwaiter awarded us a table. Bill and Matthias replayed the game from their notes. We took a cab back to the apartment over the store, where I found the corner of a yellow envelope protruding under the doorsill. It was addressed to me, and the boys read it with me:

URGENT YOU PHONE ME AT ANSONIA AT ANY HOUR
ELI

"Can you really telephone New York?" asked Matthias, all excitement at such adventure.

"I imagine so. We'll give it a try at any rate."

An operator informed me that the wires to New York were impossibly overloaded; I pleaded an emergency, and she promised the first available line. The boys wanted to stay up until the call came through, and they made a valiant effort, but Matthias was asleep by ten o'clock and Bill shortly after eleven. The phone rang at midnight. I had the receiver at my ear instantly.

"Your call to New York, Mister Kapp."

"Thank you." Immense static buzzed over the line, and then a voice.

"Sport? Can you hear me?"

"What's the matter, Eli?"

"Is that you, sport?"

"Yes, yes, what's the matter? Is something wrong with——"

"Sport, what's going on out there?"

"What do you mean?"

"What's the talk? Have you heard anything about the Series?"

"Well, of course I have, Eli! What else would people be talking about?"

"What are they saying? What are they saying about the White Sox?"

"They lost the game, Eli."

"I know they lost the game! Were you there?"

"Of course I was there."

"How did they look? Were they trying?"

"Trying! Of course they were trying, Eli. This is the World's Series."

"Are you sure?"

"Eli, what is this all about? You're not making any sense."

"Sport, listen to me. Listen carefully. I was watching the board here today. Big crowd, lots of money floating around and all kinds of rumors, too, most of them about the White Sox."

"What kind of rumors?"

"That's what I'm telling you, sport! They're saying the White Sox are taking a dive! They say the fix is in!"

"Eli, there are rumors like that every year."

"I know, sport, I know, but listen to what happened here today. Just when the game was getting started, Rothstein walked into the hotel and——"

"Who?"

"Arnold Rothstein."

"Who's Arnold Rothstein?"

"God of light, sport, were you born in a bottle? Arnold Rothstein's about the biggest gambler in the country!"

"Is he betting on the Series?"

"That's what I'm trying to tell you, sport! He didn't put any money down, just stood there watching the board until Cincy came to bat in the first. Then, when Cicotte hit the first man with the pitch——"

"Morrie Rath."

"Right, hit by a pitched ball. As soon as it happened, Rothstein left the room. After one pitch!"

"So?"

"Can you hear me? He left the room after one pitch!"

"Yes, I hear you. What's that supposed to mean?"

"I don't know, sport, but you couldn't believe what was happening an hour later. There was nothing but Cincy money in the room! Rothstein's men were covering every bet they could find, at even money!"

"They were betting on Cincinnati?"

"That's what I'm telling you, sport!"

"Was this after the fourth inning?"

"They weren't betting the game, sport, they were betting the whole Series! Cincinnati, even money! And Chicago was a two-to-one favorite just yesterday! Sport, I'll give you dollars to doughnuts that Cicotte hit Rath with that pitch on purpose. It was some kind of signal."

"Of course it was, Eli. It was a signal to the Cincinnati batters not to dig in too firmly."

"No, I think——"

"Plank hit Bresnahan with the first pitch of the 'aught-five Series."

"No, sport, this is different. I heard it from a couple of people. It's a signal that the Sox are taking a dive!"

"What people did you hear that from, Eli? Sandpaper Sam?"

"Who?"

"Eli, why don't you get hold of your friend Hal Chase. He'd know about this sort of thing, wouldn't he?"

"Hal's in California, I can't raise him. Look, sport, it's not too late. You have to go to Chicago and lay off as much as you can."

"Lay off?"

"You won't find a dime of Chicago money in Cincy, they're all betting on the home club. A bookmaker's no good, a lot of them are taking the Series off the board. You have to go to Chicago and lay off against the local money. I'll wire you a list at the Palmer House. Get down as much as you can on Cincy, sport. You've got to cover what I've got riding on the White Sox."

"How much is that, Eli?"

"About forty thousand, sport."

"Forty thousand dollars! Where did you get that kind of money?"

"My marker's good, Jackie. They know I have a piece of the company."

"My God, Eli, you haven't bet your stock!"

"It looked to be a sure thing, sport. You yourself said that the White Sox——"

"Never mind what I said! My God, Eli, you've pledged your stock!"

"I know, I know! Jackie, I'm not asking you to get ahead of this thing! I just want to lay off enough to get me even!"

"Eli, I've never made a bet in my life. Can't you get on a train——"

"I can't leave town, sport. If some people saw me trying to leave town they'd—never mind what they'd do, but I've got to stay here. I think I can get something down on the Reds, but you've got to go to Chicago."

"What am I supposed to do for money?"

"Draw it on the store's account. Think up something to tell Arthur. Please, Jackie, please!"

"If I can get the money . . ."

"Get it! I'll wire you the names at the Palmer House."

"Eli, can this Rothstein really do such a thing? Fix the World's Series?"

"The son of a bitch. I could see it if it were some Mick gambler, but to think that Rothstein would do this and not get the word to *mishpocha!*"

I didn't sleep. Instead I laid plans and devised falsehoods. In the morning I slipped out alone, for my lie needed two hours escape. It was ready when we met Arthur at Redland Field for the second game. I'd spoken to a Chicago dealer in town for the Series, I said; he'd boasted of a collection of antique stones with which he might part for a price. It was just a chance, but I was sufficiently interested to join the traffic west and examine the hoard. Mustering an offhand tone I quoted the sum of forty thousand dollars, necessarily in cash. Impossible, said Arthur; impossible to carry such an amount, he'd arrange a cashier's check. Would forty thousand be enough? And what of train reservations? They'd be difficult, but Chairman Herrmann of the National Commission had a private car, and Arthur might impose . . .

No, I said, I could put the boys on the eastbound train that evening and stay over, departing for Chicago in the morning; yes, I'd miss the third game of the Series, but that wasn't the purpose of my trip, so what matter? But I'd want to see the fourth and fifth games, Arthur guessed; he'd see to it that an aide to Herrmann delivered tickets to me. And a hotel room: he'd wire the Palmer House, although it was certainly booked to the ceiling, and use his push as a customer of a decade's standing. Would I rejoin Arthur in Cincinnati after his weekend tour of store locations? Perhaps, but I might run straight through to New York, there was no telling.

"Uncle Arthur, Daddy spoke to Uncle Eli on the telephone all the way to New York last night!" Matthias bragged.

"What did he want?" said Arthur, all attention.

I laughed. "What would Eli want? He asked how the Reds managed to win a game. I told him they outscored the White Sox. Bill, do you think Williams will have better stuff than Cicotte did yesterday?"

"I hope not," said Bill. "Come on, Cincy!"

Lefty Williams, the Chicago lefthander, had fine command save when he needed it most. He yielded just four hits, but those, and some untimely bases on balls, cost him four runs. At bat the White Sox did everything but score: ten hits garnered only a pair of runs, and those unearned when two Cincinnati throws went astray on the same seventh-inning play. To my eye that was the only suspicious moment in the game, and it cast doubt on the supposedly honest club. True, Chicago had lost the game, but there was slim evidence of fraud in collecting ten hits and yielding only four! No one who'd seen Merkle twist from the basepath and Snodgrass dribble a simple fly ball could discount the role of accident in championship play, and Eli's claim that the outcome of the Series was predetermined seemed beyond belief. I thought that he'd suffered a failing of courage, and considering the stakes I could scarcely blame him.

We moved with the mob from Redland Field to the train station; I put the boys on the eastbound sleeper, bribed a porter to watch them well, and purchased a morning ticket west. I bought a Cincinnati paper and found that the experts, who'd picked the White Sox almost to a man, were undeterred; the club which had run away with the American League pennant perhaps needed an early dousing to awaken them to the post-season challenge.

At breakfast Friday Arthur handed me a cashier's check in the amount of forty thousand dollars. "If it's no sale in Chicago, perhaps you should put some money on the Reds," he said with a smile. "They're off to a flying start, aren't they? You should have seen Garry Herrmann last night. His buttons were popping off his vest."

"Do you know, I lost the only bet I ever made?" I said. "I put a quarter down with Eli on the Giants in the Snodgrass game."

"That's probably why they lost," said Arthur. "Punishment for your sin."

I folded the check. "Arthur, a bet on Cincinnati might be safer than we think," I said. "There are some ugly rumors afloat. Have you heard them?"

"Jack, I have the absolute word of several characters that the fix is in for Chicago, and I'd believe them beyond doubt except for equal assurances by an equal number that it's in the bag for Cincinnati. Isn't that always the way? But it doesn't matter which rumors are true. What's important is that they remain mere rumors. Herrmann, Heydler, Johnson all know that they can't afford an open scandal. There have been some near things in the past, the Chase incident was one of several. The biggest names have been involved, Cobb for one, McGraw himself. Whatever's going on has to be kept under wraps. The vital thing is to hold the game together under the National Commission. Look, Jack." He took a sheath of papers from his pocket. "It's all signed, sealed, and delivered. We've exclusive rights to endorsements from every player under contract in either league. We've our pick of the crop, all for a flat fee, no more percentages. Can you imagine what this is going to be worth? We've the market to ourselves. All we have to do is produce the goods. If this doesn't bring Ruby and Saul around I can't imagine what might. There are great days ahead. Every city has its heroes, and we'll be in all of them with an officially endorsed line, locally promoted, locally manufactured, locally retailed. We'll sell franchises the way the league does, cut back our own operation to design and quality control. And with the money we make then——"

"We'll buy a ballclub," I said. "The New York Caps."

"Right. And you can pitch on Opening Day," said Arthur. "Money makes all things possible."

All things: Arthur was a devout believer. Money could purchase a massive credit to back a tenfold expansion of business, a shaded suburban hillside for a family burial ground, or an extravagant and seductive World's Series excursion calculated to curry my favor. Money might buy a ballplayer's loyalty, but it would also buy his silence, and the structure of the game would not be threatened. Money could provide a forty-thousand-dollar cashier's check on the instant. When I felt the check in my pocket I worried that the banks might be closed when I arrived in Chicago; by Monday morning the Series might be over in a Cincinnati sweep. Would the contacts on Eli's list take my marker, given the evidence of the check? What rules obtained in the world of five-figure wagering? Who held the stakes? I was out of my depth, and alone. Not for twenty years had I boarded a train in such a distracted state, and then, too, Eli had sent me forward, Chicago my destination.

The third game of the Series was played and done when I detrained, and the newsboys were crying the result, a Chicago victory. Dickie Kerr had pitched a shut-out, the White Sox had played flawlessly. My confusion grew. This might be a ploy to manipulate the odds; if so I'd best get my bets down immediately. But no team was so masterful that it could win and lose at will! I remembered Eli on the night of the Merkle game, a lost child attaching himself to every rumor that swept the Ansonia. If I followed his urging now I might rob him of a fortune. There might be a wire at the Palmer House instructing me to hold back. I taxied to the hotel, where the desk clerk greeted me with extraordinary deference; I'd stay in the manager's own suite. Money made all things possible.

There was a telegram from Eli; it listed six clients he knew to favor the White Sox and gave their phone numbers. There was also an envelope from the National Commission with Series tickets for Saturday and Sunday. Installed in the suite, I began to call the numbers in Eli's wire, but I had no answer on the first four and the fifth raised a night watchman. Business numbers! Another miscalculation. I'd have to wait until morning.

I was up at dawn; I sipped coffee and watched the clock until the instant of eight o'clock, when I made the first call. Mister Holman was not in, nor would he be that day; would I leave a message? I would not. The second call had the same result, and the third revealed that Eli's acquaintance was no longer with the firm. On the fourth call I reached the contact, a Mister High. He knew my name; he admired my work. How was my brother? No, not the younger one, good old Kappy; he'd missed him that summer.

"He's busy at home and couldn't make the trip," I said. "Between the two of us we've seen every Series game since 'aught-five, and I had to come out to keep the record intact."

"Do you follow the game like Kappy does?"

"Exactly," I said. There was a pause.

"Who do you like?" he said.

"I'm a National Leaguer from 'way back. I like the team that Mathewson built."

"Oh, you do. Well, they started out strong, but now we've got them at home and it's a different story."

"I don't think so," I said.

"Oh, you don't. Well, how strongly do you disagree?"

"Pretty strongly."

"Do you like Cincinnati today?"

"I hate to predict one game or another. I've a pretty strong feeling on the whole Series, though."

"Oh, you do. Well, so do I."

"I thought you might. That's why I called."

"Oh, I see. Well, could you put a figure on your feeling?"

"I can go pretty high," I said.

"How high is pretty high?"

"Up to forty or so."

"Up to four?"

"Forty."

"Oh, forty! Well, that's a lot of money."

"Kappy said you might be in that league."

"Oh, he did. Well, not by myself, but I've got some friends."

"I thought you might."

"What are you asking?"

"Beg pardon?"

"What are you asking? What odds?"

"Oh. Even money, I suppose."

"You're up, two games to one."

"The Sox were two-to-one three days ago."

"True enough. Even money, you say. Well, it's tough to find any Cincinnati money in this town at any odds. Even the books won't handle the action."

"You've found it here," I said.

"So it seems. Well, I tell you what. Let me make some calls around and I'll see what I can put together. When can we meet?"

"It's up to you," I said.

"Will you be at the park?"

"I plan on it."

"Tough to meet at the park, though, in that mob."

"I'll be here until nearly game time."

"Where's here?"

"The Palmer House."

"Oh, the Palmer House. Well, listen, forty thousand's a lot of money. It might take a little time. Here's what I'll do. I'll call you before the game and give you a figure, and then we can meet afterwards and exchange markers. You've got cash to back them?"

"I've a cashier's check I can exchange Monday morning."

"Why not today?"

"Are the banks open? Mine doesn't do business on Saturday."

"Mine does. Go to the First National, mention my name."

"I'll do that."

"Listen, your marker would be jake with me, but there'll be other people who like to see the color of the coin."

"Fine with me."

"I've got plans for dinner and a show," he said. "Clients in town. Care to join us?"

"Sorry, I've plans myself."

"Oh, you do. Well, then, why don't I meet you at your hotel around midnight?"

"I'll be here."

"Look, I'll give you ten—no, fifteen thousand right now, and I'll call you around noon to let you know about my friends."

"I'll go to the bank and be back by noon," I said.

"First National. Mention my name."

"I'll do that."

"So we've got fifteen down for sure, and anything up to forty, at even money."

"That's right," I said. "You won't back out if Cincy wins today, will you?"

"Say, my word's my bond. Check with Kappy if you don't believe it. These bets are made."

"If your word's good with Kappy, it's good with me."

"Fine. And now I'll tell you a little secret," he said, lowering his voice.

"Go ahead."

"It's the ball," he said.

"It's what?"

"It's the ball. They used a National League ball in Cincinnati, Cicotte didn't like the feel of it. Neither did Lefty Williams. Out here they're using the American League ball, that's why Dickie threw a shutout yesterday. That's why Cicotte will get 'em today. It doesn't matter if the Reds win all the games in Cincinnati. They've only got two more back there, and it takes five to win."

"I hadn't thought of that," I said.

"You've got to know the inside dope. You still want the bet?"

"I'll chance it," I said. "I'll expect your call around noon."

Four stacks of hundred-dollar bills, a hundred bills to a stack: forty thousand dollars cash proved amazingly compact. My hand never lost touch with the money as I walked back to the hotel, while my mind did calculations. Forty thousand dollars was ten years' salary to me. I more than matched it in dividends and special stipends, but I could live in comfort on the salary alone. Forty thousand dollars was a month's pay for a whole division of buck privates in the late war. It could cover the entire White Sox payroll, for Cicotte himself made no more than five thousand, nor did Jackson or Weaver. Forty thousand dollars was nearly as much as Mathewson had earned in salary in his career. Yet with a few phone calls my newfound friend might match the sum; little wonder Chicago had been Eli's favorite stop on the swing.

The phone rang at noon; I started, calmed myself, and answered it.

"Did you get the money?" the voice asked.

"I did."

"Oh, you did. Well, you can put thirty-five thousand aside for our purposes. There are takers for today's game if you want to risk the other five."

"No, I'm only interested in betting the whole Series."

"Oh, you are. Well, thirty-five's as much of your money as we can take."

"Regardless of today's game," I said.

"These are solid bets, my friend. Did you check with Kappy?"

"No need for that. I don't doubt you."

"You're a wise man. I'll look you up at midnight."

"I'll be here."

"Like I say, your handshake would be jake with me, but these other fellows——"

"I understand."

"Oh, you do. Well, good. See you tonight."

In mid-afternoon I joined the pilgrimage to Comiskey Park. Ignited by the White Sox victory the day before, the faithful packed the twin decks of the grandstand in historic number. They bore the ornaments of their commitment: buttons of every size and slogan, noisemakers of assorted dimension, folded streamers of blue and white to fling into the autumn air and punctuate a White Sox hit or a Cicotte strikeout. These were midwestern faces bright with anticipation, men and women together, wrapped in scarves and sweaters of varied hue to ward off the chill of the incessant wind that blew in from the lake. The people of Chicago knew their baseball and loved their heroes. How generous that all the fervor so lately leashed to war could flow out over the green acres of a playing field, inspiring young men to action in a nobler competition! See the knights who bear the city's emblem on their breasts: they are all in white, their lances are burnished wood, their gauntlets calf-brown leather. Watch them race over the manicured lawn, see their silver spikes strike fire! See their tan faces in shadow beneath the bills of their caps, watch them laugh, observe the steel in their glances as they assay the gray-clad opposition. It is all but impossible to envision them in defeat; it is beyond imagination that they would seek it out.

And now look away from the field, look up, and see the men who chart and chronicle the event, see them set apart from the great host, these makers of heroes and legends: see Lardner, Runyon, Rice; see Taylor Spink in a place of honor as keeper of the sacred scrolls, the *Sporting News;* see young Pegler throw sheets of yellow paper at the harassed telegrapher. And see Fullerton carefully set his pencils and his scorebook before him, see him wipe his rimless spectacles with a white handkerchief. Now he rises and performs a slight deferential bow to the figure who slides into the seat beside him; look at the gaunt cheeks and great bent hands, and know that he alone among those jurists has performed on the field below, to win such glory and to suffer such devastating disappointment. See Christy Mathewson on high.

I tried to watch the game as he might, but taken together all the multiple simplicities of play overwhelmed my amateur eye. How could I tell Cicotte's true intent when he allowed Rath a lead-off single, then induced a ground ball for a double-play? Did Groh's pop fly to shortstop signal the pitcher's strength or the batter's weakness? Neither team threatened in the first inning, and Cicotte disposed of the Reds easily in the second, striking out the last man. In the bottom half Joe Jackson hit an unthreatening fly ball to right-center. I watched the hitter run, or rather trot, toward first, but then a great shout went up and Jackson put his head down and raced around the bag. I looked to the outfield and saw Roush in pursuit of the bouncing ball. In studying Jackson I'd missed the play in right and learned it only when I heard the word "misjudged" bandied about. Jackson was bunted to third, the execution flawless; Risberg walked and stole second without a throw; Schalk drew an intentional pass. Cicotte could have struck out without raising a doubt about himself, but he hit a sharp grounder to second. It was an out to end the inning, but a worthy try.

In the third all the evidence went against Cincinnati. Wingo dared Schalk's arm, the best in baseball, and was out stealing in the top half, and in the bottom Jimmy Ring, the Reds' pitcher, hit a batter on the arm, and then Rath bobbled a grounder for an error. Two men on, but Felsch grounded out and another zero went up on the board. Now the fourth, a critical inning in the White Sox losses, but no damage was done either way: six batters, six outs. I looked to the pressbox and saw Mathewson sitting upright, his attention on the field, while Fullerton hunched over his scoresheet, the scribe at his duty.

Edd Roush led off the Cincinnati fifth. The National League's best batter had but one hit in the Series while his counterpart, Jackson, had six; was this, and Roush's second-inning error, grounds for suspicion? It was difficult to remember that the White Sox were the reputed villains, especially when Roush only nubbed the ball a few feet in front of the plate, where Schalk pounced on it and threw the man out.

Pat Duncan was the man most likely to be forgotten when reciting the Cincinnati lineup. He hadn't much of a bat and still less of an arm; his war record better explained his place on the roster than his talent, and he was starting only because a veteran regular was hurt. Cicotte worked him with curve balls, and the third was belt high. Duncan blasted it back at the pitcher, Cicotte's glove went up, and the ball hit leather and soared into the air above the mound. Cicotte looked about, heard or felt or sensed the ball bounce behind him, stabbed at it and whipped a hard throw to first. Had the throw been true it would have been a very near call, but it shot through the tangle of runner and fielder at first and bounded up the line, with Schalk in pursuit. Duncan glanced about and took off for second, narrowly beating Schalk's throw.

It was a mistake, Cicotte's mistake, but the result of hard play and a sporting effort. Cincinnati's mistakes afield were far less excusable.

Now Larry Kopf, the thin Cincinnati shortstop who'd shocked his teammates, and perhaps himself, with a triple in the second game, his longest hit of the season. The pitcher kept the ball low and away, twice for balls, twice for strikes. The fifth pitch came in the same place, and Kopf reached for it and stroked it on a line into left field. Here was all the intricate movement of a play in baseball: Duncan taking tentative steps toward third, then exploding into a full gallop as he saw that the ball would drop in front of Jackson; the left fielder dashing to the line of the hit, overrunning it slightly to ensure the freedom of his gloved left hand as he scooped the ball from the grass; shortstop Risberg aligning himself and raising his glove to offer Jackson a near target for his direct throw home; the third baseman to his station, the second baseman to his, and the catcher, Schalk, stripping the mask from his head and flinging it far away as Duncan turned third and headed home. And the forgotten portion of the play: Kopf, the batter, rounding first and making for the extra base that a play at the plate would allow him.

Cicotte didn't forget. I couldn't tell if Jackson's throw home was true—Schalk seemed to be leaning to his right in anticipation—but Cicotte dared to intercept it, granting the run for the chance to cut down Kopf at second. The ball glanced off his snug mitt and bounced where no fielder was stationed, the deserted space between home and first. So neither runner was tagged, neither out was made, and the play ended as it had begun, with a man on second and one out, but the score was one to nothing.

Greasy Neale came to bat. His nickname was far more memorable than his talent, but any man with a bat can do damage, especially when he goes against his book. He took Cicotte's pitch to the opposite field, Jackson's field, and Joe was positioned close behind shortstop; he gave chase, but the ball fell behind him and the wheels spun again, coming up as before: a man on second, one out, and the score Cincinnati two, Chicago naught.

One can find bad business in heaven if the search is undertaken with a predetermined will. The Cicotte plays—the throw to first after Duncan's smash, the cut-off attempt to catch Kopf at second—seemed the result of excessive effort, not its lack. I might have taken the chance at Kopf had I thought of it; I hadn't, but I was a fan, no veteran of fifteen campaigns. If one takes a chance one risks an error, and those with the best fielding averages are often the most timid men; moreover, the official scorer had marked neither play an error. As for Jackson, if he played Neale in the same spot a thousand times he'd be right all but once. Bad business, bad judgment—I couldn't tell, and I turned to the pressbox to observe the one whose judgment I'd trust, but Mathewson was gone. Fullerton remained, busy with his notations, but the pitcher was nowhere to be seen.

At first I looked for him after every pitch as Cicotte shut down the threat, then after every batter as the game progressed. Mathewson did not return. Groh made another error for Cincinnati and Roush remained hitless, but Jimmy Ring—a Brooklyn boy, I remembered, and not much older than Bill—pitched to make the old neighborhood proud. Gandil's sixth-inning single was Chicago's last hit. A threatening sky darkened the field, reflecting the gloom of the home crowd, and isolated knots of brave Ohioans provided the only animation in the park. It ended two to nothing; the stands were soon empty and silent, save for the ring of the telegrapher's bell and the skipping keys of the pressbox typists.

I hailed Fullerton when I saw him close his scorebook. He squinted, removed his glasses and returned my wave. I shouted that I wanted to see him; he cupped his hand around his ear, shook his head, and gestured that I should sit and wait; he'd come to me. I walked down an aisle to the first row of seats. I heard shouts from the dugout and saw a team of groundskeepers rush onto the field, their aspect a parody of the gallants who'd played the game. They swarmed over the roll of tarpaulin directly in front of me and cursed and sweated to push it away from the rail. A blast of thunder answered their groans. I felt the rain, and then I felt it no more; Fullerton's umbrella was shielding me.

"What do you make of it, Kapp?" he said.

"Quite a storm," I said, watching lightning strike over the lake.

"The heavens weep for Chicago," he said. "What can I do for you?"

"Why did he leave in the fifth inning?"

"Why not ask him yourself?"

"Would he see me?"

Fullerton looked thoughtful. "When he learned you'd arranged a Cincinnati club ring for him, he wept. Of course he weeps easily these days."

"He must have reason," I said.

"He does. He's a dying man, Kapp, but he doesn't weep for himself. There's a great deal else."

"Tell me."

"I will. Come on, I'll drive you to the Congress Hotel."

It was a Ford Model A. Fullerton revved the motor and backed out of the space, turning north. The rain beat steadily on the roof as Fullerton peered through the windshield.

"Do you know he's under treatment, Kapp?" Fullerton asked.

"At Saranac," I said. "I gather it was an accident. I know he was responsible for disposing of poison gas."

"Yes, he'd barely marched off the boat when the Armistice was signed, but he was determined to do dangerous duty. A whole strip of the Lowlands is thick with unexploded shells and canisters, Kapp. Farmers will be plowing up mines for a generation. Think of it. Killed in the war, twenty years after it's over."

"I appreciate the irony."

"It's worth an essay, but I'm a mere sportswriter. It seems that Matty had an untoward encounter with some mustard gas, in Flanders field, poetically enough. A tragic accident, if accident it was. Opinion differs."

"We asked the War Department if a medal was due him. They said not."

"My sources told me that when you've buried a million heroes sympathy lags for a living one, all the more so when evidence suggests that he initiated the incident. Well, then, you know all that. Let's come to Cincinnati."

"Have you asked him what happened?"

"Kapp, in my circle it's considered impolitic to mention a rope in the house of a hanged man. Let's come to Cincinnati." We stopped at a traffic signal, and Fullerton cleared the fogged windshield with his white handkerchief. "I was surprised when

Pulitzer announced that Matty would be covering the Series for their papers. He turned down an offer from my syndicate in August. I'm sure that if it were any other team but Cincinnati in the Series he'd have stayed at Saranac. When I read the Pulitzer release I wrote to him and arranged to share a room in Cincinnati. He agreed, and I met him there."

"How did he look?"

"Hollow. His body retains its width and breadth, but it's as if all the muscle and sinew has been sucked out of him."

"And his mind?"

Fullerton shook his head. "Kapp, a dying man has a realm of thought the living never enter. These thoughts intrude on his mind from time to time. He gives them vent, as if no one were listening, and then he comes back to himself. One makes allowances. Damn it, let me tell you what happened in Cincy!" His anger surprised me, and I held my questions.

"I told him there were rumors. That's all I said, just that, and he flashed at me. 'Don't say that about my club!' he shouted. 'Don't even breathe of it! I cleaned out my club, there's not a man on it I wouldn't trust with my life!' He hardly moved, but he had me backing away with my hands raised in surrender. I said no, the rumors weren't about his club, they touched the White Sox. Of course there are always rumors about the Sox, it's a Chicago tradition. We have the gentlemanly Cubs on the North Side and the rowdy Sox down here, it's been that way for a generation. It's been worse these past few years, though. There were the White Sox, world champions in 'seventeen and hardly losing a man to the war, but they folded last season and this year they dropped quite a few games they ought to have won. Half the club is living beyond its salary—well, there are vagrants on State Street who live beyond the salary Comiskey pays, but that just fuels the matter. If any club could be bought it's the White Sox.

"Kapp, tip me on a horse race and I'm down for a double sawbuck. Tell me a fighter's going into the tank and I'll believe it. This job could turn Francis of Assisi into a butcher. But the thought of half the club conspiring to throw a World's Series! And——"

"Half the club!"

"And the rumors were floating on the other side as well. You must have heard them, Kapp. Every name on either roster has been spoken with a wink and a nod. Yes, it happens every year, but now the racing crowd's at the ballpark, and the odds were shifting suspiciously. I wired my accounts to hold off their bets. Better safe than sorry, and all that. The question is, how can you tell? It's a subtle game, Kapp, and we've both seen men we'd trust with the crown jewels play like circus jugglers. Who's to judge?"

"Mathewson."

Fullerton nodded. "Yes, Mathewson. I asked him to sit with me in the pressbox, and if he saw a play that aroused his doubt, any doubt at all, he was to tell me and I'd circle it on my scoresheet."

"What has he seen?"

"The book's right here. Have a look."

The scorebook was the size of a large sketchpad, bound with wire. I turned to the last page and worked backwards to find the current games. There they were, in the book's middle portion; the markings were sharp.

"First game, fourth inning," said Fullerton. "One on, grounder to Cicotte. Double-play ball if it's handled cleanly, but it wasn't. Cicotte's throw to second was off the mark. They got the out there but lost the man at first." I found the square easily; it was circled in red. "Cicotte skipped and double-pumped on the throw. It looked like he was making sure, but of what? Matty didn't like it. And then the next play, the infield single to shortstop. Risberg backed up on it. Matty quoted McGraw: 'The first step backwards is the one that kills you.' Then the single to right and Shano Collins' throw home. Hopeless. It allowed Wingo to go to second."

"Slack play all around," I said.

"Yes, and these are the White Sox! Defense is their strong suit. Three red circles, but Matty cautioned me that they were only possibilities. Now check the second game."

I turned the page. "Look at Williams pitching in the fourth," said Fullerton. "Lefty's a control artist. He averaged fewer than two walks a game all year. Now here are three in a single inning. And again a bad throw home from the outfield on a base hit, it let Groh take third."

I looked at the red circles. "I've never known a game without mistakes," I said. "When two fine teams have at it the mistakes are decisive. Games are more often lost than won."

"That's the devil of it, Kapp. Every foul-up has a thousand precedents, that's the bad man's camouflage. Skip the third game, there's nothing there. Go to today's game."

There were circles around a second-inning play and two fifth-inning squares, the last boldly marked.

"Risberg's stolen base?" I asked.

"Second inning, right? Men on first and third? All Risberg does by stealing is take the bat out of Schalk's hands. First base is open, they walk Schalk, and Cicotte's up with the bases loaded. No reason for it. And then in the fifth, Cicotte's throw to first on the comebacker by Duncan." Fullerton steered the car through a turn. "The important thing on a bang-bang play like that one isn't the out, it's keeping the man at first. If you have to hurry that throw you oughtn't make it at all. And then Cicotte's cutoff play on Kopf's single."

"I thought it was smart baseball, worth a try," I said.

Fullerton pulled to the right lane and slowed the car as we approached the Congress Hotel. "I didn't draw that mark," he said. "Matty did. Then he left the pressbox."

I looked at the thick red circle. "What made him so sure?" I said.

"Ask him yourself. Here we are."

I folded the scorebook and watched the rain pelt down on the sidewalk. "Are you going to the National Commission with this?" I said.

"I've been turned down flat," said Fullerton angrily. "It's all politics. Comiskey and Ban Johnson aren't speaking. Herrmann has conflicting interests as Commission president and Cincinnati's owner. His club's winning the Series, isn't it? And each league would love to see the other wrapped in scandal. Do you suppose McGraw would weep for Ban Johnson? Besides, they've buried so much they're sure they can keep the lid on this one." He hit the steering wheel with his palm. "Their contempt is enormous, Kapp. It encompasses the players, their chattel and weapons against one another. It includes the writers, who they think they can buy with a sirloin steak and two fingers of whiskey—and in nine cases out of ten they're dead right. And, of course, the fans. Crooked or honest, the Commission runs the only game in town, and they know it. Jesus, Kapp, you can't find a dozen real gamblers who think the Series is on the level, yet they lay their money down and pray that they're on the right side of the fix. It's a stinking situation. I'll dig into it as best I can, but—to quote my editor, who won't let me run any of this—opinion isn't proof and rumors aren't evidence."

He fell silent. I pulled the handle but held the door shut against the rain. "Are you coming up?" I asked.

Fullerton turned to face me. "I was with Matty when Herrmann told him you'd made a Cincy ring in his name. He felt it deeply; the very mention of your name changed his mood. When Herrmann was gone I took the moment to ask him about you. I recalled our encounter in Boston and wondered how he could call you a friend when you'd hardly known one another.

" 'But I never called him a friend,' Matty said. 'I called him a celebrant. He was a voice of annunciation many years ago, and the time will come for another encounter. He'll not deny me.'

"Matty was in quite a state when he left the pressbox today, Kapp, but I think this is your time. You'll find him in Room Three-thirty-six."

He was wearing the ring of the three diamonds, but so dark was the room that the stones were dull on the platinum band. The heavy drapes were drawn against the weather, and a small bedside lamp cast its light on a Gideon Bible beneath it. He was wrapped in a varsity sweater with an N and Y intertwined in the Giants' logogram. His hair was unkempt, as if the rain had spoiled it. There were thick folds and furrows on his forehead, and his lips and jaws moved as if he were swallowing bile. He sat on the edge of the bed with his stockinged feet on the carpet, and in his huge, bent hand he held a large gray handkerchief which he used to cover his soft, persistent coughing.

He spoke first of my daughter. "She was in my prayers, as were you," he said. "I'd have written, but other affairs weighed upon me. I apologize."

I shook my head.

"I've often thought it might be better to have had a daughter," he said. "Unfair expectations ensnare my son whenever he picks up a glove. I encourage his play, of course, but he oughtn't think of a career. Does your boy play?"

"On occasion. There aren't the opportunities. The places I used to play are all built over."

"Is he here with you?"

"No. I had him in Cincinnati, but he's home now."

"He was at your brother's great supper? I'd have liked to meet him, but my strength must be husbanded. I wonder what the game means to him? Does it attract him as it did us, when we were young? Or is his life all straight lines and clear decisions? Has he no need to escape into the game?"

"He studied the war news more closely than the sports," I said.

"Maps and toy soldiers? My boy as well. His heroes wear a different uniform than Amos Rusie did. Or does now, taking tickets at the Polo Grounds."

"I know Matthias enjoys his outings at the park with his grandfather."

"He'll remember that when the day comes that he takes his own grandchild," said Mathewson. "Do you recall your first game?"

"Oh, yes. It was at Prospect Park in Brooklyn. Eli and I happened upon it. We were——"

"How is Eli?" he said, interrupting. "Come, tell me."

"He's retired. He has no connection with the firm."

"I know that. Heydler's report had a letter to that effect from your younger brother. He's expert at letters as well as telegrams. But Eli's well?"

I felt the money in my pocket. "At the moment, no. He's backing the White Sox heavily in the Series, and he fears he's on the wrong side."

A gruesome smile crossed Mathewson's face. "I wonder what he made of Cicotte's play today," he said. "I wonder if it told him what it told us. He's here, in Chicago?"

"No, he's following the play at the Ansonia. Rumors are running rampant there. I had a call from him when I was in Cincinnati. It's the reason I've come here, to Chicago." As I talked,

Mathewson leaned back onto the bed and, with some effort, stretched his legs out over the bedcovers. He closed his eyes, but gestured with his hand that I should continue.

"He's pledged his stock in the firm to back the bets. If he loses it will leave him bereft. He begged me to come here and lay an equal sum on the Reds."

"Have you done it?"

"I've made some calls. I've yet to exchange notes." I waited for him to say something, anything, but he was silent. "I must know if these rumors are true," I said. "If the Series were honest I'd let Eli fall of his own misjudgment, but if he's caught up in foul play——"

"No, no, no, no, no," he said. "That isn't why you've come, not to have me confirm what you already know. The game's as open to you as it is to me."

"You do me too much credit," I said. "You've always done me too much credit. I'm not an educated man. The classic quotations in your wires were mysteries to me, I needed my wife to interpret them. My knowledge of the game is rusted by twenty years' disuse. I was never more than a sandlot player, in any case."

"You might have been."

"No, never. And not for an unsound arm, either. That's an excuse of Eli's devising. The truth of the matter was that my parents wouldn't allow me to sign a contract. I couldn't—I wouldn't go against their command. I hadn't your courage."

"Nor my willful disobedience. And here I lie." He looked at me. "I tell you again that you need none of my wisdom to see the truth of the matter in this World's Series. It takes no profound knowledge of the game to know where a pitcher belongs on an outfielder's throw home. You must have learned that, even on the sandlots."

I saw the play, Cicotte interrupting Jackson's throw twenty yards in front of home. "Back of the plate," I said. "He should be backing up the catcher. He had no business in the middle of the diamond."

"Indeed not," said Mathewson. "Indeed not." He brought the handkerchief to his mouth, and when he took it away it was speckled with blood. "So, you have your answer. If that's the sole reason you've come here, you may go now to exchange your notes and rescue your brother." He stared at me, daring me to rise and depart. Nothing could have made me do it. He knew it; he closed his eyes and laid his head back on the pillow.

"It isn't Eli who stands on the precipice," he said. "It's you, you, who sways there. It's you that risks damnation. Do you see that? How can I teach you? Come, there's a box in that drawer. Will you bring it to me?"

It was an oblong leather box, light in my grip. I handed it to him, and he struggled to sit upright, then placed the box on his lap and opened it.

"Here, the ruby," he said, lifting the first of the seven rings that lay within. "St. Louis, July fifteenth, Nineteen and One. The fire burns in the stone, do you see? So long ago, yet the fire still burns. And this, its sister, the emerald. Almost black in this poor light, but in the sunshine how it sparkles! Here is the club ring for the championship in 'aught-four, our first. I have your sketch for it, remember? What pride I took in wearing it! And the next year's championship—you didn't like the piece, it was done to satisfy McGraw, and so you crafted this one, the one I wear. This diamond trinity."

Three rings remained in the box, and he ran his fingers over the velvet on which they rested. "Nothing for 'aught-eight, of course," he said. "Poor Fred Merkle, how could he know that he was a mere pawn to be sacrificed in the middle game? And I was as ignorant as he. We are not given to understand until we have reached the end." He coughed again and daubed at his lips with the handkerchief, then lifted another ring. "There was the rain in 'eleven, a week of it, a tortuous time. And then the awful trial of the following year, a trial I failed. Yes, I failed it, distrusting my teammates and disgracing myself, paying the price when Speaker's foul ball fell untouched. To blame Fred Snodgrass, and then to allow that! I wasn't worthy of the prize. And this, the last pennant we won in New York. But not the world's championship, we were decimated with injury and denied ultimate victory. And it was just.

"We grew old together in the seasons following. Our eyes would meet at a fatal point of play, we'd see our age and know we were fated to lose. We had to endure those dreadful harangues while Mac, as young as his youngest player, repurchased his youth by renewing his roster. And again I failed to understand. I imagined that Mac's part was mine, that I'd regain the pinnacle as he would, with a team under my command. Hal Chase taught me the error of that.

"Yet once more I mistook the lesson. I thought my fate was to die in war, under the banner of my country. I sought it out, but even there I was denied; when I awoke in the hospital I wept to be alive. What was my purpose? Would I come in the end to Rusie's station, a totem in a gray usher's suit?"

"Never," I said.

"Of course not. Both of us know I had a greater role to play. It drew me to Cincinnati. My doctors told me I oughtn't go, but I knew I must, and yes, there was Fullerton with his dark and ugly tale, insisting that I alone could tell the truth of it.

"I feared it. I feared the truth I saw on the field, and far more I feared the truth I realized in myself. Yet how could I deny it? It was as clear as the fire that burns here, in the stone of your first offering. I'd walked among those men on the field, I was as they were yet far better than they could ever be. I'd achieved the perfection you celebrated in stone. Then followed doubt, confusion, failure, and finally betrayal; then followed my death, for it was death, in the explosion and the pain. And then I rose from that death, I walked among the people as of old, and finally, finally, I came to sit in judgment of those I'd walked among, to root out their sin and damn them for it."

I drew my chair closer and leaned toward him, struggling to hear his whispered words. "I do damn them," he said. "With a mark I damn them. I damn Cicotte. I damn Jackson. I damn Risberg and Gandil and Williams. And if there be others I will damn them as well, I will root them out and damn them for eternity. And I damn the filth that corrupted them, the dicers and the high rollers. They will pay. They will pay in time. I shall not rouse them now, for I will allow them their full portion of loss, and when the corrupters are counting their gains I shall spring upon them and drive them from the temple!"

With that his head fell back onto the pillow and his hand came down among the rings. Some fell to the floor and others scattered about his body; one remained beneath his hand. It was the ruby, and he lifted it and placed it on his finger.

"How can I doubt this when you are here to witness it?" he said softly. "You have always been with me as my fate has uncoiled. You were there at the first, you knew even then how high I stood above other men, you announced my advent with this ring, you celebrated my glory with another and another, and you marked my faltering steps with the rest. Do you still doubt what you must do now? Do you, my celebrant?

"Look at me. Look into my eyes. Think of this: they diced for His robe while He suffered on the cross. Will you do that, while I lay dying? No, you will not. No matter what may follow, you will not do that. Not you," he whispered. "Not you."

He lay still. I covered him with white blankets; I collected the rings and placed them in their velvet settings. I held the leather box in my hands as I sat on the hard chair by the bed, and I waited for him to sleep. After a time I placed the box inside the Bible under the lamp, snapped out the light, and left him.

I wept as I walked from the Congress to the Palmer House in the driving storm. In the face of the desk clerk I saw the sight I must have made, but it mattered nothing. I asked if there were a late departure for New York, and when he said I had an hour to make the last train I told him to draw my bill. I took a sheaf of hotel stationery and scrawled a note:

ALL BETS OFF.

I handed it to the clerk with the instruction that the message should be given to a Mister High when he called for me at midnight.

I never slept all the forty hours to New York. Before me was the image of guiltless Mathewson in agony. I saw the handkerchief blotted with his blood, the yellowing parchment face whispering damnation. If I thought of Eli, it was to guess at what portion of my income I must donate to his support when his fortune was gone to the gamblers. I filled a sketchpad with fantastical drawings on themes of death and resurrection, crumpling each effort in my fist and throwing it on the floor before launching another.

I went directly from Grand Central to the Ansonia, where I found Eli gloomily watching the odds. I told him I'd not placed the covering bets, and he must suffer his loss; he wailed that he must sell his stock to Arthur, and I agreed he must, and gave him the forty thousand dollars cash against his benefit from the sale. My pledge of support did not comfort him. Pale and trembling, he turned his back and walked out of the hotel onto Broadway.

The gamblers never saw Eli's money. Much of it arrived in the next morning's mail at the synagogue which had been the scene of all our frequent mourning. The balance was spent at the Chalmers Automotive Agency that Eli visited an hour after he left the Ansonia. The salesman later swore that Eli had seemed excited but not unduly distraught; most people got rather excited when they purchased a Chalmers, he said. Eli selected a glorious showroom model equipped with every late gadget and gewgaw. The car was noticed; people were found who recalled that it entered Central Park at Columbus Circle and grandly toured the inner roadway, heading north. Our investigator reported that the Chalmers exited on upper Fifth Avenue near the site of Bennett's old polo field and headed west along the northern tier of the park. It climbed Morningside Heights and turned right at the Cathedral, continuing north on Amsterdam. At One Hundred Fifty-Fifth Street it turned east and increased its speed; it was at full throttle when it jumped the roadway onto Coogan's Bluff; it hardly slowed as it jounced over the thin grass and solid rock; it sailed high into the air over the cliff and described a full somersault before it hit the jagged cliffside and tumbled over and over and over again until it came to rest, shattered and burning, against the black walls of the Polo Grounds.

October 7, 1925—at Pittsburgh

										R	H	E
WASHINGTON	0	1	0	0	2	0	0	0	1	4	8	1
PITTSBURGH	0	0	0	0	1	0	0	0	0	1	5	0

BATTERIES: Washington, Johnson & Ruel;
Pittsburgh, Meadows, Morrison (9) and Smith, Gooch (9).
Winning pitcher: Johnson. *Losing pitcher*: Morrison.

THE WORLD'S SERIES

	W	L	PCT.
Washington (A.L.)	1	0	1.000
Pittsburgh (N.L.)	0	1	.000

Washington leads best-of-seven series, one game to none.

EPILOGUE

MATHEWSON was seven years dying. Word of his death rode a midnight wind through the hotels and speakeasies of Pittsburgh on the eve of the World's Series of 1925; both teams played the next day with black bands of mourning on the sleeves of their uniforms. The obituaries were full and flowery, and mentioned only in passing his term as Cincinnati's manager. There was wonder when his will revealed his wish to be buried there.

The Black Sox scandal had broken onto the front pages in September of 1920. With Mathewson secluded at Saranac Lake, Fullerton pursued the story alone; he was pilloried by the baseball establishment, and his own newspaper refused to publish much of what he wrote. At last an Illinois prosecutor with his eye on the governor's race convened a grand jury and called the players to testify. Cicotte admitted to several instances of dishonest play, as did others. Each incident was circled in red on Fullerton's scorecard, so accurate was Mathewson in his seat of judgment. Eight players were tried. They were defended by William Fallon, a flamboyant New York attorney with ties to McGraw. His argument was inventive: not that play had been honest, but that no existing statute had been broken. On that basis the eight were acquitted. Nevertheless they were done for, as was the National Commission. Kenesaw Mountain Landis, a Federal judge with a stern aspect and a reputation for integrity, was installed at the head of a new professional structure, and he barred the offenders for life. He also canceled the Commission's contract with Collegiate Jewelers.

McGraw left Pittsburgh to attend the funeral. His skein of four pennants had ended that year. In 1921 and '22 the entire Series had been played at the Polo Grounds, and with Ross Youngs outshining Babe Ruth the Giants won both times. In '23 the Yankees had a park of their own, larger than the Polo Grounds and within its sight on the Harlem River; Ruth hit three home runs in that Series, and the world's championship crossed to the Bronx. McGraw never won another. In '24 a bad bounce on a ground ball cost him a seventh-game victory, and then the team declined; Ross Youngs sickened, the wonderful body withered with disease, and he died in '27, the year Ruth hit sixty home runs. For five years thereafter the Giants were close in the race, but with never enough to win. In '32 McGraw quit. Twenty months later he was dead.

Before he died there was a final act. In the summer of '33 the promoters brought the best of both leagues together at the Chicago World's Fair; they called the clubs the All-Stars, and they brought out McGraw to manage the Nationals. Connie Mack ran the Americans, with all those big fellows: Ruth, Gehrig, Foxx, Simmons, Cochrane. When the managers shook hands at home plate the mind raced back nearly thirty years, and the talk was all of Mathewson and the three shut-outs.

McGraw's team lost, to a home run by Babe Ruth.

I didn't weep at Mathewson's death. It was an escape from the body that tortured him, and it raised him beyond the heights he'd known in youth. He became legend.

I had a package from the executor of his will. It contained the sketch I'd done and signed for him in 1904, and I hung it beside the drawings of the ruby ring and the triple diamond. The accompanying note told me that the rings had been buried with him, the triple diamond on his finger and the others in the leather box at his feet. It was then that I wept, but on the night he died I thought, where is King Kelly? Dead these thirty years. Where is Big Ed Delahanty? Did they ever find the body, when they dragged the river below Niagara Falls? Where is Cap Anson? Where's Monte Ward? Charlie Ganzel, Buck Ewing, Big Dan McGann? All gone; the flesh is mortal. The legends abide.

The legends abide. According to the baseball records that were young Bill's passion, one Edward Sylvester Nolan entered the National League with the Indianapolis club in 1878, two years before Mathewson's birth, three before my own. Nolan is a common enough name, and it was coincidence that from his own time to

McGraw's and beyond he was the only Nolan to play on a major league roster; from that coincidence his nickname derived. He was known then and forever as The Only Nolan. His five seasons were spent with five clubs in three different leagues. He led Indianapolis in walks that rookie year, and he lost far more often than he won, but he pitched a shut-out, a rare feat in those days and the only one of his career. A boy watched him pitch that game and was inspired by it. I do not know where the inspiration led him or what it finally cost him in family and fortune, but I met him as a man one evening, and we talked baseball. Records and statistics meant nothing to him; the greatest pitcher he ever saw was The Only Nolan. The greatest I ever saw was Christy Mathewson, on a terribly hot day in St. Louis, young as an April morning in that sweltering July, the perfect pitcher. It was my happiness to celebrate that perfection; in his age and suffering he would accept that vision of my youth, entwine it with his own hard faith, and end in madness. Eli, Eli!

ACKNOWLEDGMENTS

THE PLAYERS, managers, owners, and league officials in this novel are historical figures, and the ballgames described herein were played on the dates given; line scores, batting orders, and the standings of the clubs are accurately reproduced. Very few liberties have been taken with the play-by-play, and none at all where World Series games are concerned. All off-the-field incidents and conversations are fictitious, and the brothers Kapp, their family, and their associates are entirely imaginary.

The first edition of the *Baseball Encyclopaedia* (Macmillan, 1969) is the source for the player's ages, home towns, and yearly statistics and for the final standings of the clubs, season by season. Mid-season standings are taken from the *New York Times* of the appropriate date, and the "Censurable stupidity on the part of player Merkle" story ran in the *Times* of September 24, 1908. Other citations from newspapers of the day are fictitious.

The World Series (Dial, 1979) describes in close detail the play in post-season competition, and the author owes a special debt to Richard Cohen and David Neft, who compiled it. Details of other games have been reconstructed from reportage in the *New York Times* and from the reminiscences of participants, especially as recorded in Lawrence Ritter's *The Glory of Their Times* (Macmillan, 1966). Eliot Asinof's *Eight Men Out: The Black Sox and the 1919 World Series* (Holt, Rinehart & Winston, 1963) and Joseph Durso's *The Days of Mister McGraw* (Prentice-Hall, 1969) excited the author's imagination, and Mr. Durso was gracious in answering further inquiries from the author. The New York Public Library provided important information about

New York in Mathewson's time. Knowledgeable readers may find inaccuracies; the author hopes that these will not diminish their pleasure in this work of fiction.

New York City
November 1979–February 1982